VACUUM MICROBALANCE TECHNIQUES

VOLUME 2

W0071836

VACUUM MICROBALANCE TECHNIQUES

VOLUME 2

Proceedings of the 1961 Conference
Held at the National Bureau of Standards,
Washington, D. C., April 20-21

Edited by R. F. Walker

High Temperature Reactions Group
National Bureau of Standards

With an Introduction by
H. S. Peiser

Metrology Division
National Bureau of Standards

Springer Science+Business Media, LLC
1962

Library of Congress Catalog Card Number 61-8595

ISBN 978-1-4899-6155-6   ISBN 978-1-4899-6285-0 (eBook)
DOI 10.1007/978-1-4899-6285-0
Softcover reprint of the hardcover 1st edition 1962

CONFERENCE PARTICIPANTS

Behrndt, Klaus H. Autonetics
12000 East Washington Blvd.
Whittier, California

Bertino, James P. Los Alamos Scientific Laboratory
P. O. Box 1663
Los Alamos, New Mexico

Blackburn, Paul E. Westinghouse Research Laboratories
Churchill Borough
Pittsburgh 35, Pennsylvania

Boggs, William E. U.S. Steel Corporation
Applied Research Laboratory
Monroeville, Pennsylvania

Cahn, Lee Cahn Instrument Co.
14511 Paramount Blvd.
Paramount, California

Chappel, Frank A. L. Oertling Ltd.
Cray Valley Works
St. Mary Cray, Orpington
Kent, England

Cochran, C. Norman Alcoa Research Laboratories
Box 722
New Kensington, Pennsylvania

Czanderna, Alvin W. Union Carbide Research Labs.
Parma Research Center
P.O. Box 6116
Cleveland 1, Ohio

Domingues, Louis P. Bureau of Mines
College Park, Maryland

Donihee, James B. Bureau of Mines
College Park, Maryland

Gulbransen, Earl A. Westinghouse Research Laboratories
Pittsburgh 35, Pennsylvania

Haliday, William B. Electron Devices Section
National Bureau of Standards
Washington 25, D.C.

Hagel, W. C. Metallurgy and Ceramics
General Electric Research Labs.
P.O. 1088
Schenectady, New York

Hampson, R. F. High Temperature Reactions Group
National Bureau of Standards
Washington 25, D.C.

Katz, M. J. U.S. Army Signal Research and
Development Laboratory
Fort Monmouth, New Jersey

Kern, Sandford Lincoln Laboratory
C-310, MIT
P.O. Box 73
Lexington 73, Massachusetts

Knapp, Edmund C.   Shawinigan Resins Corporation
Springfield, Massachusetts

van Lier, J. A.   Research Laboratory
Union Carbide Consumer Products Co.
P.O. Box 6116
Cleveland 1, Ohio

Little, James W.   National Bureau of Standards
Washington 25, D.C.

Macurdy, L. B.   Mass and Scale Section
National Bureau of Standards
Washington 25, D.C.

McNish, A. G.   Metrology Division
National Bureau of Standards
Washington 25, D.C.

Madorsky, S. L.   Polymer Structure Section
National Bureau of Standards
Washington 25, D.C.

Paule, Robert C.   Department of Chemistry
University of Wisconsin
Madison 6, Wisconsin

Paulson, Rolf A.   Analytical and Inorganic Chemistry
Division
National Bureau of Standards
Washington 25, D.C.

Peiser, H. Steffen   Consultant, Metrology Division
National Bureau of Standards
Washington 25, D.C.

Rodriguez, Mrs. Patricia M.   Raytheon Company
Research Division
Waltham 54, Massachusetts

Rouse, Glenn F.   Electron Devices Section
National Bureau of Standards
Washington 25, D.C.

Russell, T. W.   Quinco
Boulder, Colorado

Stockbridge, C. D.   Bell Telephone Laboratories
Whippany, New Jersey

Volpe, N.   Wm. Ainsworth and Sons
Boonton, New Jersey

Wade, William H.   University of Texas
Austin, Texas

Walker, R. F.   High Temperature Reactions Group
National Bureau of Standards
Washington 25, D.C.

Warner, A. W.   Bell Telephone Laboratories
Whippany, New Jersey

Wolsky, S. P.   Raytheon Company
Research Division
Waltham 54, Massachusetts

Zdanuk, Edward J.   Raytheon Company
Research Division
Waltham 54, Massachusetts

EDITOR'S NOTE

This volume contains the majority of the papers presented at the 1961 Conference on Vacuum Microbalance Techniques, which was held at the National Bureau of Standards, Washington, D. C., on April 20-21, 1961.

The 1961 Conference was a direct outgrowth of the first Conference, held at the U.S. Army Signal Research and Development Laboratory in 1960, and of the enthusiasm engendered at the first Conference for a continuing series of such meetings. In planning the second Conference a conscious effort was made to obtain the participation of a wider representation of the investigators throughout the world interested in microweighing. It was perhaps inevitable that the effort was more successful in terms of the wider range of techniques used by the investigators than in terms of international participation. Nevertheless, it was gratifying to have one paper presented from England and to receive several letters of inquiry and interest from other European countries.

It is to be hoped that the trend to broader participation will continue and that the Conferences will continue to stimulate the exchange of information on the science of measuring small mass differences. Provisional agreement of participants has been obtained to hold a third

Conference in the West Coast area during the fall of 1962, and to hold a fourth Conference in the Pittsburgh area in the spring of 1963.

On behalf of the participants appreciation is expressed for the Conference facilities provided by the National Bureau of Standards. Thanks are due in particular to Dr. E. Wichers, Associate Director of the Bureau, for his welcoming remarks, to the various staff members who assisted in arranging laboratory visits, and to the authors of the papers, who reduced to minimum proportions the task of editing their manuscripts.

R. F. Walker

November, 1961

CONTENTS

Contents

INTRODUCTION

On the subject of conferences many a scientist will express a strong personal preference for the small, highly specialized meeting attended by scientists actively engaged in research closely related to the relevant speciality. Constructive intercommunication with others seems simpler, faster, and happier than at larger meetings. The conference series on Vacuum Microbalance Techniques is a perfect example of gatherings of active scientists with a common interest. The reader may sense from these pages that this second conference of the series was indeed an outstanding success.

It is true that at an organizational session there was a searching discussion on whether in the attempt to keep it small and informal the meeting had been publicized inadequately. Probably all of us could think of people who would have enriched the conference discussions.

At the same session plausible arguments were put forward both for increasing and for further restricting the scope of the series. As an example, most members felt F. A. Chappell's contribution was most appropriate and worthwhile—an ultramicrobalance built on the accumulated experience of several of the most famous names in the ultramicrobalance art—yet, strictly, it was not a vacuum instrument. The instrument, however, clearly could operate in a moderate vacuum by the trivial addition of a "bell jar." Discussion of this and similar balance designs was therefore considered relevant.

The conference listened enthralled to the papers of Stockbridge and Warner, and of Wade and Slutsky, on mass measurements with resonating quartz crystals. Yet the equipments described are remote from micro-balances of general applicability. The concensus of opinions nevertheless favored the inclusion of this type of paper.

This brought the session to the inevitable discussion of definitions. Future conferences should discuss definitions more fully. One might ask for example whether most of the instruments discussed are not best termed ultramicrobalances, a definition for which, I think, should include two conditions:

1) Small sensitivity (say 10^{-7} g per smallest readable unit).
2) Very small mass of the beam and hang-down assembly.

The first four papers of this volume describe microbalance systems which should be considered in conjunction with the instruments described and referred to in the first volume. The advances are really spectacular especially, I think, because the operation of most of these balances does not have to be left exclusively to experimentalists of great skill. Modern microbalances are rugged instruments that are not easily damaged. In many ways they are less easily disturbed by changes in ambient conditions than are conventional chemical balances.

Most of the remaining papers presented at the conference dealt with applications to high-temperature studies, vapor pressure measurements, chemical reaction rates, and measurements on thin films. Without further introductory remarks they amply testify to the increasing usefulness of vacuum microbalances providing experimental techniques of diverse applicability.

L. B. Macurdy's contribution in a sense differs from all others as it discusses the calibration of mass standards for microbalances by direct comparisons ultimately descended from the International Kilogram. These mass standards provide the means by which the precision inherent in microbalances can be converted to accuracy. Buoyancy and chemical methods can, it is true, be used effectively for calibration. In the hands of some they may in some instances provide superior accuracy at present. Nevertheless, in indirect methods the sources of uncertainty introduced are probably harder to recognize and assess than in direct mass comparison series. Macurdy and others working in this field ought to be encouraged to devise means of providing very small accurately calibrated reference masses of proved mass constancy.

Perhaps I may be permitted to make one more general comment on direct calibration methods. The uncertainty in the calibrated value of a small weight depends on only four terms:

1) The uncertainty in the mass value of the reference weight. This uncertainty is involved only through its product with the mass ratio of the unknown to the reference standard. In other words, if we use a 100-mg reference weight to calibrate a 10-mg standard, only a tenth of the uncertainty in the former is passed on to the latter.

2) The uncertainty in the mass value of the balance sensitivity weight. If the calibrating balance linearity is checked and the weights are closely

adjusted to integral multiples or submultiples of each other, only a small fraction of this uncertainty is passed on to the calibrated weight.

3) The precision of the calibrating balance. (Usually the largest term for weights substantially smaller than the reference standard.) The observational equations in the calibration series provide several degrees of freedom. An estimate of the effect of the dispersion of the measurements on the value of the weight is obtained from the coefficients of the observational equations.

4) The mass constancy of the calibrated weight. Macurdy's paper properly points out the importance of this term which is controlled and should be watched by the user.

One of the features of the conference series appears to be the location of the meetings near centers of interest in ultramicrobalances. It is most appropriate therefore that a meeting near the several prominent Californian laboratories is planned. In this way, too, we hope that the second conference at the National Bureau of Standards was enriched by laboratory visits where the instruments described in this and the previous volume could be subjected to critical inspection. Conference members were also able to see experimental models that might develop into contributions at future meetings. A 2-g quartz torsion microbalance of very simple design with a sensitivity of about 10^{-7} g was especially notable. Mr. T. W. Russell of Quinco built this instrument after discussions with us.

I conclude my introductory remarks with the conviction that this little volume will make a significant contribution to the literature on microbalances. I also hope readers of this book will be stimulated to join in contributing to this rapidly advancing field of research. At future conferences of the series, they will, I am sure, be warmly welcomed.

H. S. Peiser

November, 1961

THE CAHN GRAM ELECTROBALANCE

Lee Cahn and Harold R. Schultz
Cahn Instrument Company
Paramount, California, U.S.A.

ABSTRACT

Previous Cahn Electrobalances have been and are widely used in vacuum weighing. Their principal limitation has been capacity. A new Electrobalance has been developed using an elastic ribbon suspension, with capacity in excess of 1 g. Precision is 0.1 μg. The new instrument also has new circuitry to permit extremely long-duration runs at full precision without adjustment, and provision for taring to increase percentage precision, up to 2 ppm. The new instrument can be used in high vacuum, and with samples which are at high temperatures. It is not affected by vibration, normal shock, temperature, or temperature changes.

INTRODUCTION

Electromagnetic balances are particularly well suited to vacuum micro-weighing. Remote control is easy; the only connections through the vacuum seal are electrical. It is easy to produce arbitrarily small force increments. Electrical measurements can be made to finer fractional

7

accuracy and precision than is generally possible with deflection or Hooke's law balances.

A simplified diagram of our first electromagnetic microbalance is shown in Fig. 1. It was originally developed for spectroscopic and biochemical use, but immediately found wide use in vacuum weighing as well. The beam and coil go into the vacuum, while the electrical controls remain outside for manipulation by the operator. The beam is made of aluminum, and is counterweighted for the popular "one pan" type of operation. While aluminum has a much higher temperature coefficient of expansion than quartz, it also conducts heat much better, and thermal gradients are much smaller. Even with severely asymmetrical heating from the lamp, the stability of the zero remains within the sensitivity of the instrument for weeks at a time. Pivot and jewel bearings are used for the central bearing, and wire bearings for the stirrup. This is a true balance, with free stirrup bearing, so that readings are independent of position of

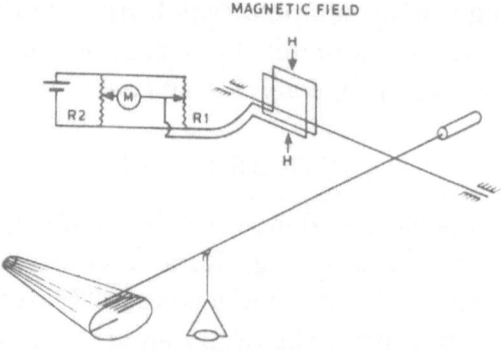

Fig. 1. Simplified diagram of original electro-magnetic balance.

sample on the pan, and total weight values are accurate.

The moment of the unknown sample is balanced by adjusting R1 until the beam is balanced. The current in the coil is exactly proportional to the electromagnetic torque, and thus to the sample weight. This current is measured by potentiometer R2, which has a ten-turn dial on its shaft. Auxiliary circuits, not shown, subtract tare weight and calibrate R2 to read directly in milligrams or micrograms. Calibration is done against standard weights and eliminates the effects of temperature, component drift, level, etc. Various ranges can be obtained by changing the resistance in series with the coil (not shown).

Ultimate precision of this design is about 0.3 μg, which is the standard deviation of a single reading determined by removing and replacing the test weight, with 15 mg of stirrup and pan. Sensitivity, determined by noting the minimum reliably detectable change in weight of a sample which remains on the pan, is better than this, since the shock of removing and replacing the sample contributes a major component of the error.

Fractional precision is about 0.02% of full scale on any range, and is determined by the resolution of the electrical components. Accuracy is determined only by the linearity of R2. We have never detected any nonlinearity in the torque-current characteristic up to full-scale ranges of 0–50 mg. Up to 0.03% deviation from linearity may occur on the highest range, 0–100 mg. R2 is normally a 0.05% potentiometer. Capacity is rated at 175 mg, although higher loads can be used with less precision.

The principal limitation of such an instrument for

vacuum work is the need for calibration as the battery drifts. It is not possible to operate at full fractional precision for longer than 8 hours, and even this requires some care. Nevertheless, we estimate that more than 100 such instruments are used in vacuum, for thermo-gravimetry, corrosion research, etc.

This limitation was overcome in our next design, which is self-balancing, automatic recording, completely line-operated and temperature-compensated. It uses the same torque motor used in the previous design. A simplified diagram is shown in Fig. 2 and a photograph in Fig. 3. Deviation of the beam from balance is detected by the phototube. The phototube current is amplified and applied to the torque motor to rebalance the beam. The torque motor voltage represents the sample weight to about 1 part in 10,000. It is compared with a reference voltage from a potentiometer calibrated in milligrams, and the difference is applied to a recorder. Four ranges are provided on the reference or suppression potentiometer:

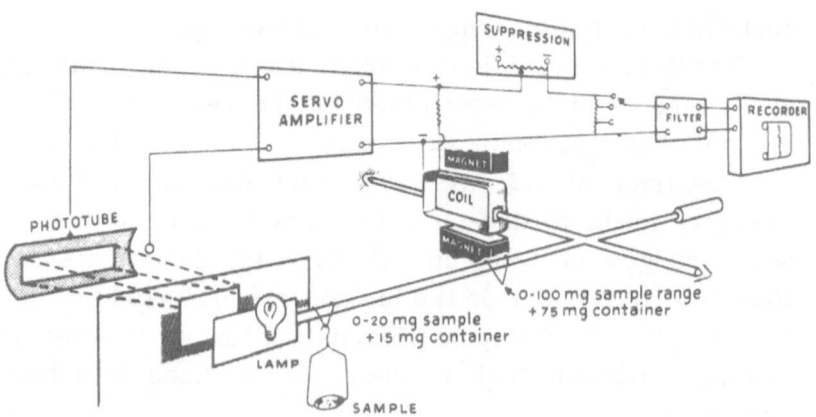

Fig. 2. Simplified diagram of automatic recording electromagnetic balance.

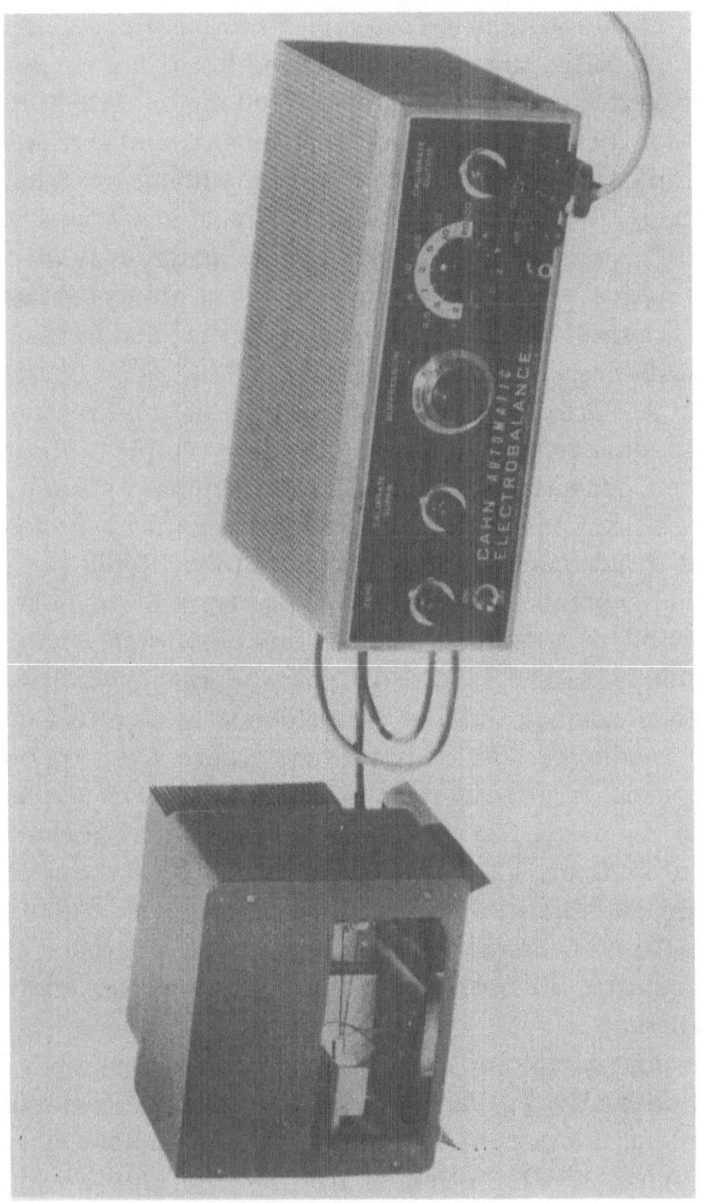

Fig. 3. Automatic recording electromagnetic balance. For vacuum use the weighing mechanism is removed from the smaller case, and installed in the vacuum. Beam, phototube, and lamp are all aligned to an aluminum bracket, and all leads go to a connector. Adaptation to glass tubing or bell jar is simple.

0–10, 0–20, 0–50, and 0–100 mg. Fractional precision is 0.01% of full scale on each range. Recorder ranges are provided from 0–0.2 mg to 0–100 mg in 1-2-5-10 steps. Both recorder range and suppression dial settings may be adjusted at will during a run, without breaking the vacuum. Thus it is possible to measure a weight change in vacuum to within 0.01% of the change. It is also possible to scale the instrument to present either coarse or fine changes full scale on the recorder, and to vary the scale during a run.

Ultimate precision is about 0.3 μg on consecutive samples; sensitivity on a single changing sample is about 0.2 μg. An exceptionally stable reference power supply and careful temperature compensation permit operation at rated fractional precision indefinitely. Rated performance is obtained with line voltage from 65 to 140 v. It is possible to adjust the temperature coefficient during production to within ±0.002%/°C over a range of about 10° C. Thus normal ambient variations have no effect on balance readings. The temperature range over which compensation is effective can be changed electrically in a simple manner. The torque motor has been operated successfully from −20° C to above 100° C. It is possible to obtain extremely high sample temperatures without exceeding 100° C at the balance beam, with a furnace of normal quality. We have never heard of a need for additional baffling.

This instrument has been pumped down to, and operated successfully in, vacuums to 10^{-8} torr. It remains an excellent instrument for many vacuum applications.

Its principal limitation is its capacity. Also, even higher percentage precision would be desirable in cor-

rosion and surface chemistry work, and even finer ultimate precision is always good. It was obvious to us that
these objectives required an entirely new approach to
torque motor design.

THE GRAM ELECTROBALANCE

Our new torque motor uses an elastic ribbon suspension, as illustrated in Fig. 4, in place of the pivot
and jewel type. The ribbon is made of a beryllium copper
alloy, 0.015 mm thick, and will support more than 5 g.
More than 100 of these suspensions have been tested for
hysteresis, and we have not been able to detect any evidence of it. This is attributed to the small angle through
which the beam rotates in this type of balance, 4° total
spread. This is entirely different from the problems
involved in a torsion-type balance, in which the elastic
member is stressed through a much wider angle.

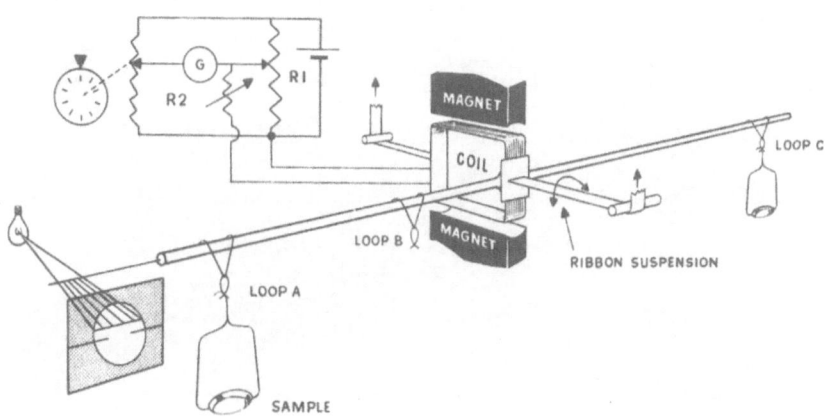

Fig. 4. Simplified diagram of Gram Electrobalance.

No nonlinearity has been observed on any range. A potentiometer linear to 0.05% is normally used, but provision has been made for a potentiometer having 0.01% linearity if needed.

The wire bearings have also been revised, as shown in Fig. 4. This instrument also uses a different configuration, the old familiar "two pan." By taring the container or part of the sample weight, it is possible to

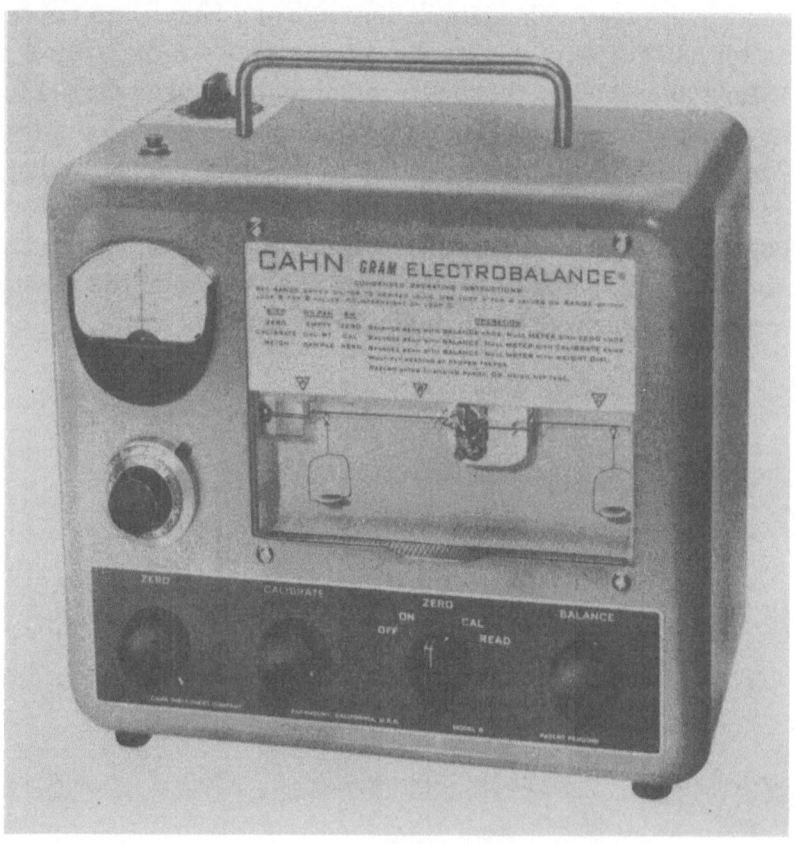

Fig. 5. Gram Electrobalance.

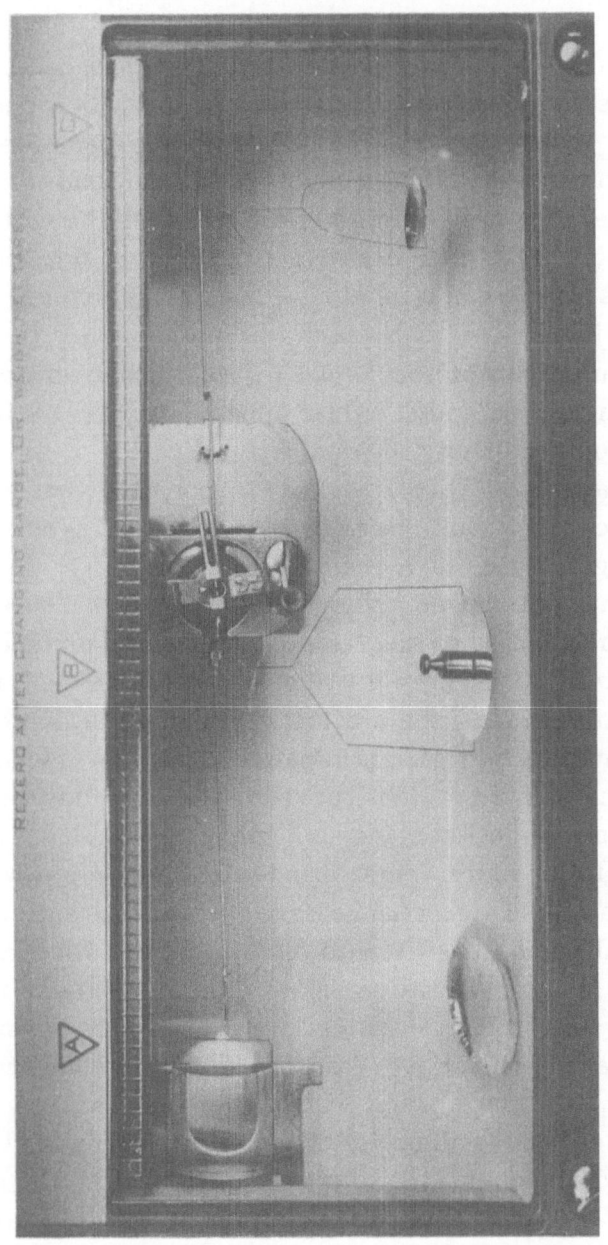

Fig. 6. Detail of weighing assembly. The torque motor is removed for vacuum work.

obtain a precision of 1 part in 10^5 of total load with con-secutive samples, or 2 parts per million with a changing sample which is not removed from the pan. Since the precision is definitely a function of total load, for most work there is no advantage in exceeding the conservative 1.5 g capacity limit. Above approximately 5 g load the beam is arrested by a stop, preventing further stress on the ribbon.

Ultimate precision is 0.1 μg, for either consecutive or changing samples. This appears to be limited by deflection sensitivity. Optical magnification of $2\times$ is used to achieve this level, and it is possible that greater magnification would permit even greater precision on changing samples in vacuum.

Electrical ranges are provided from 0–1 mg to 0–1 g. Refinement of the electrical components extends the electrical precision to 0.01% of full scale on each range. In this instrument, the electrical range would be set only to cover the c h a n g e anticipated in vacuum, rather than the total sample weight, as with the previous models. Thus one might use the 0–1 mg range with a 100-mg sample to measure weight changes of up to 1 mg to 0.2 μg. If the sample weight change were greater than anticipated, one could simply switch to the 0–5 or 0–10 mg ranges, without losing calibration. The precision of these ranges is also 0.01% of full scale. In general, it is possible to measure to almost 0.01% of the change without breaking vacuum, for any change within the total capacity of the instrument, regardless of whether this change was an-ticipated or not. It is also possible to reverse the direc-tion of the electromagnetic torque at the control unit, without losing calibration, for samples which lose weight.

These features are not possible with other types of balances.

With deflection-type beam balances, the deflection sensitivity, and thus the reading, varies with total load. This effect is especially pronounced with the type of balances commonly used in vacuum micro-weighing. The electromagnetic balance described here does not have this problem, since the beam is always returned to a reference position. Its readings are thus not affected by the total load, even though its deflection sensitivity is.

The instrument is made with a line power supply, which will remain within 0.01% for supply voltages from 95-130 v. It uses cascaded zener diodes and no vacuum tubes. Long-term stability is exceptionally good. Temperature compensation is also used in this instrument to minimize the effects of ambient temperature. It can also be adjusted to cover temperature ranges other than ambient. Thus long runs, up to at least several weeks, are possible.

The materials used in the design were chosen to give it the same or better high-vacuum and high-temperature capabilities as the previous models.

It is thus well-suited to all types of vacuum weighing, including many applications which the previous models could not satisfy.

All of the instruments described above are quite insensitive to vibration, and may be used in areas where vacuum pumps and other sources of vibration are operating. They are always shipped fully assembled and ready to use, with less than 0.3% incidence of reported shipping damage. Under extreme vibration the only change in performance may be a decline from the "weight change" performance levels to the "consecutive sample" level. As noted above, they are also unaffected by normal ambient temperature changes, by extremes of temperature, and by level and vacuum. We have not heard of any damage due to controlled atmospheres in which they have been used, which include O_2, H_2, and others. They are also convenient to operate, since the condition of beam balance is easily observed, without the use of cathetometers, and they read directly in milligrams or micrograms, without any calibration factors.

The elastic ribbon suspension would also appear to offer major advantages in a recording balance. We have constructed a prototype of such an instrument, but cannot yet report performance values in vacuum.

CONCLUSION

An electromagnetic balance with elastic ribbon suspension has been developed, which offers substantial advantages in vacuum weighing. It is controlled remotely. Rated capacity is 1.5 g. Sensitivity is 2 parts per million of load, while weight changes can be measured to 0.01% of the change. It is unaffected by vibration, level, and normal ambient temperature changes. It can be used in high vacuum and to weigh samples which are at high temperatures. Readings are not affected by total load and are obtained directly in milligrams or micrograms.

THE DESIGN AND CONSTRUCTION OF A NEW QUARTZ FIBER BALANCE*

F. A. Chappell
L. Oertling Ltd., Cray Valley Works, St. Mary Cray, Orpington, Kent, England

ABSTRACT

The problems concerned in the commercial production of a quartz fiber ultramicrobalance are considered in relation to a number of existing designs. It is concluded that none of these lend themselves to this type of production, and a design finally adopted is described.

Performance tests show that the balance enables weighings to be carried out with a standard deviation of 0.08 μg, up to a maximum load of 250 mg in each pan. The factors influencing the accuracy attainable are discussed.

INTRODUCTION

Recent developments in the use of ultramicroanalysis led L. Oertling Ltd. in 1956 to investigate the extension of their series of micro-chemical balances into the sub-micro range, i.e., the range of instruments capable of

*The design of this balance is the subject of patent applications, in the United States and other countries.

weighing to 0.1 μg or better. The following is an account of the considerations which have determined the final design of a quartz fiber ultramicrobalance, together with typical performance data.

BEAM DESIGN REQUIREMENTS

Since the start of this century, many descriptions of beam balances with high sensitivities have been published, and several authors have reviewed these designs.[1,2,3] In recent years, fused quartz has been adopted as the constructional material for the beam, since it offers the advantages of low thermal expansion, low moisture adsorption, and a high strength/weight characteristic. The form of pivot for the beam and pan suspensions has also received considerable attention.

Many authors have pointed out the limitations of knife edge bearings when used in balances of this type, and the best performance appears to have been obtained when flexible suspensions were used. If quartz is used for these suspensions, its freedom from elastic hysteresis ensures that the suspension is virtually free of friction. A disadvantage of this form of construction, however, is that it is impossible to define precisely the position at which the flexible suspension starts to bend, since considerable distortions can occur close to the joint, due to the fusing process. This means that both the arm length and the alignment of the end suspensions with respect to the central pivot have to be corrected by a subsequent bending of the beam. This operation can only be carried out after the weighing performance of the beam has been determined, and is, therefore, a most unsatisfactory feature if the instrument is to be made in large numbers.

It is worthwhile pointing out here that although some variation in optical sensitivity* with load is permissible in an instrument which is to be operated by a null method, no variation can be allowed where the deflection of the beam is to be used as an indication of mass, unless calibration is carried out at each load. Even with null-type instruments in which the beam is restored to a fixed position by a controlled torque, a low optical sensitivity can impair accuracy owing to reading errors, and a high optical sensitivity is associated with a long period of swing, making the operation of the instrument tedious.

Using the dimensions of the beam used in the new Oertling balance, we can calculate the effect of an error in the vertical position of the attachments of the pan suspension fibers by means of the formulae derived by Carmichael[2] and Cunningham.[1] It can be shown that, if the attachments are $10\,\mu$ high, the optical sensitivity at a load of 200 mg in each pan is 2.6 times the value at no load. If they are $10\,\mu$ low, the optical sensitivity drops to approximately 60% of its no load value. Figure 1 shows the effect graphically and includes the types of suspension used by Carmichael (central torsion fiber and vertical pan suspensions) and Day[4] (central torsion fiber, vertical beam suspension, and vertical pan suspensions). Also shown is the variation of sensitivity with load on the new Oertling beam. In each case, it has been assumed that the no-load sensitivity has been adjusted to a predetermined value. It is obvious that, if the optical sensitivity is to be kept reasonably constant throughout the

*Defined as the ratio between the angle of rotation of the beam and the mass causing it.

F. A. Chappell

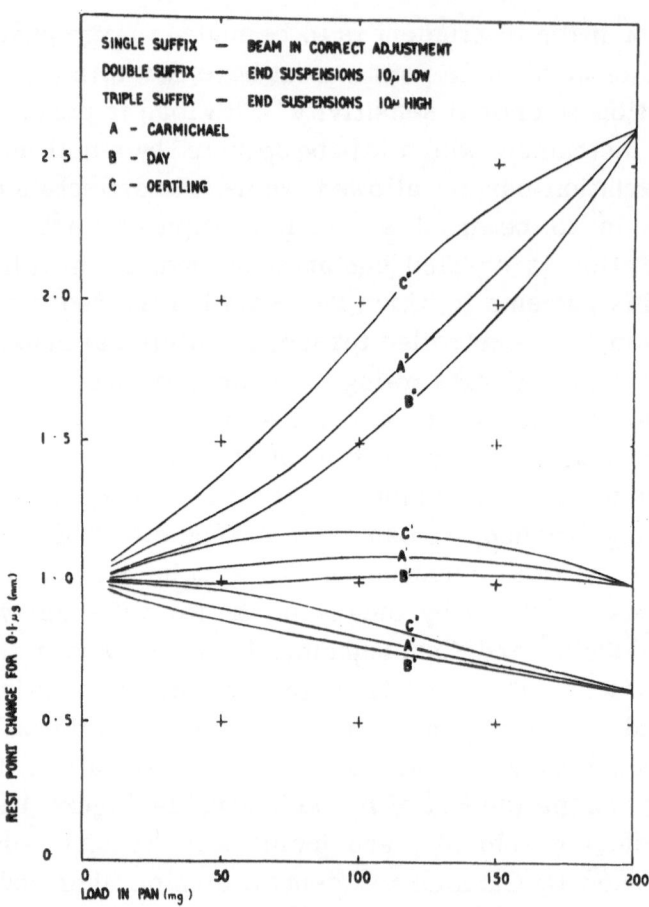

Fig. 1. Variation of optical sensitivity with load.

load range of the instrument, control of the vertical positions of the pan suspension fibers is necessary to within a few microns, and this is quite impossible with the Carmichael and Day designs. The effect of unequal arm length due to inaccurate horizontal positioning of the pan suspension fibers is well known to all users of conventional balances.

Since some variation of optical sensitivity with load can be expected, the use of a null technique in weighing eliminates the necessity for calibration at each load, and the provision of a controlled restoring torque with a wide range reduces the use of very small weights. The horizontal quartz torsion fiber is ideal for this purpose, since it has a linear relationship between torque and rotation, and in addition the fine fibers required can be turned through several revolutions. By using a fixed horizontal fiber colinear with the torsion fiber at the other side of the beam, a form of central pivot for the beam is provided. Unfortunately, this form of suspension sags under load, and the vertical position of the beam varies with the weight in each pan. This makes determination of the null position of the beam difficult and necessitates a double optical system. The use of a vertical beam suspension fiber eliminates this difficulty, and at the same time increases the load capacity of the instrument. In this case, the fixed horizontal fiber only serves to keep some slight tension in the torsion fiber. A further advantage of the vertical beam suspension is that, since the torsion fiber does not carry any load, the torsion head sensitivity is completely independent of load.

From the above it will be seen that the design requirements for the beam of an ultramicrobalance to be manufactured in quantity are as follows:

(a) Beam to be made of fused quartz.

(b) Flexible suspensions to be used for the beam and pan suspension pivots.

(c) Position of the suspensions to be accurate to within a few microns.

(d) Weight of the beam and pans to be supported by vertical suspensions on the beam.

(e) A horizontal quartz torsion fiber to be used to restore the beam to its null position.

(f) A rear horizontal fiber to keep the torsion fiber in slight tension.

BEAM DESIGN

The form of beam construction finally adopted is shown in Fig. 2. The main structural members are two 300-μ-diam quartz rods A, which lie side by side and are spaced by the pointer B and the two tie bars C and D. The pointer and tie bars are of 200-μ-diam quartz rod. Attached to the two rods A are three transverse cylinders E, F, and G of 200-μ rod. These cylinders form the locations for the beam suspension fibers H, pan suspension fibers J, torsion fiber K, and rear fiber L. Their cylindrical form enables them to be located precisely during assembly by a manufacturing jig. The beam and pan suspension fibers are led around the cylinder and fixed at the opposite side. The position at which the suspensions bend is thus determined by the cylinders themselves, and not by the fixing. It will also be seen that, as the beam rotates, the increased bending of one pair of suspensions is counteracted by the reduced bending of the other pair. Consequently, the vertical suspensions do not oppose movement of the beam. A quartz spring N makes it possible to keep the tension in the torsion fiber at a preset value.

With this design the addition of a mass m to the right-hand pan at arm length L produces an angular beam movement θ, which is balanced by restoring forces

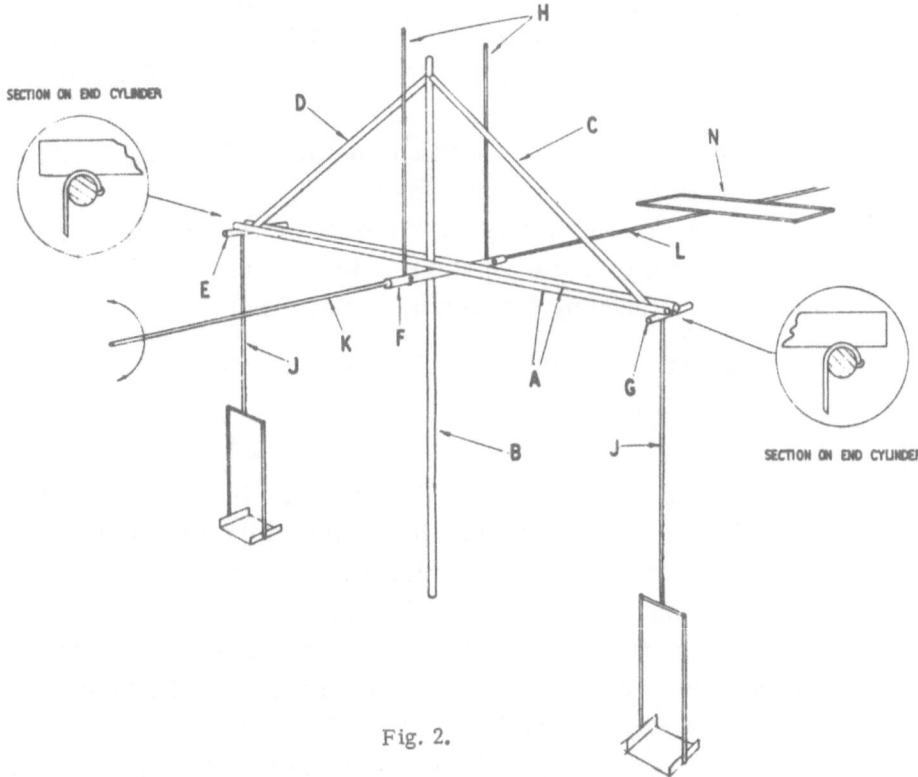

Fig. 2.

$M_b g d\theta$, due to the mass of the beam; $2Mg(h_0 + Mk)\theta$, due to the axis of rotation of the beam being above the line joining the end suspensions; and $2(n_t\pi r_t^4/2l)\theta$, due to the rotation of the ends of the torsion fiber and the rear fiber.

    M_b is the mass of the beam

      g is the acceleration due to gravity

      d is the distance of the center of gravity of the beam below the axis of the center transverse cylinder

M is the load on each pan suspension

h_0 is the height of the axis of the central transverse cylinder above the line joining the axes of the end cylinders with no load on the pan suspensions

k is a proportionality constant relating to bending of the beam under load

n_t is the modulus of rigidity of the torsion fibers

r_t is the radius of the torsion fibers

l is the length of each torsion fiber

The equation for the optical sensitivity is thus

$$S_0 \equiv \frac{\theta}{m} = \frac{L}{(n_t \pi r_t^4 / g l) + M_b d + 2M h_0 + 2M^2 k}$$

It can be seen that if the beam is made perfectly rigid, i.e., $k = 0$ and the three cylinders arranged to be in line, i.e., $h_0 = 0$, the optical sensitivity is completely independent of load. It will be noted that there are fewer terms in the denominator of this equation than in the formulae giving the optical sensitivity of the beam designs of Carmichael or Day. It follows that a larger value of d is required on this beam to give a particular sensitivity. In fact, d is approximately 80 μ, compared with a figure of about 20 μ for a Carmichael beam with the same specification. With the Day beam, it would be necessary to make d about -70 μ to overcome the extra restraint of the beam suspension fibers.

The low position of the center of gravity of the beam is undoubtedly one of the major factors contributing to the balance's great stability.

Although techniques for fusing fine quartz fibers and rods together have been developed,[5,6] the necessity for very precise control of flame size and temperature, and the problems of flame position make this method more suitable for a laboratory than for a precision instrument factory.

However, Wilson and El-Badry[7] and Asbury, Belcher, and West[3] have described balances in which the quartz is cemented together. It should also be remembered that on the Carmichael design of beam, the attachment of the torsion fiber to the torsion head—one of the most critical joints on the whole beam—is by means of a cement.

The quartz rods and fibers comprising the beam, pans, and suspensions of the new balance are cemented together, and no difficulties in manufacture have been experienced, apart from the necessity of ensuring that the amount of cement at each joint is as small as possible. Adjustment of the position of the center of gravity of the beam is also carried out by means of small amounts of cement positioned as required.

TORSION HEAD SENSITIVITY

The torsion head sensitivity, i.e., the ratio between the angle of rotation of the torsion fiber ϕ and the mass m which has to be put on the pan to restore the beam to its original position, is given by

$$mgL = \phi \frac{n_t \pi r_t^4}{2l}$$

Therefore

$$S_T \equiv \frac{\phi}{m} = \frac{2lgL}{n_t \pi r_t^4}$$

In this instrument, $l = 2\frac{1}{16}$ in., and, taking the value of n_t as being $3.7 \cdot 10^{11}$ dynes/cm^2, the diameter of torsion fiber required is $21.86\,\mu$.

MECHANICAL DESIGN

The construction of the instrument is shown in Fig. 3. To make possible immediate replacement of the beam in case of damage, it is attached to a support by means of the beam suspension fibers and torsion fibers. This support also carries the final gear in the torsion head drive and can be lifted out of the instrument when two screws are removed. Beam replacement, therefore, consists of lifting off the cover of the instrument, removing two screws, and withdrawing the damaged beam complete with its support. A new beam unit can then be put into position and the screws and lid replaced. It will be noted that the cantilever design of the beam support ensures that the tension in the torsion fiber is not changed when the beam unit is fitted to the balance.

The torsion drive consists of antibacklash gears and worms and gives a total reduction of 20 : 1. Since five complete revolutions of the torsion fiber are used in covering the torsion head range, the operating shaft and knob turn through 100 revolutions. An engraved drum on the shaft has 100 divisions, each division being approximately 3 mm wide. Three counting wheels driven from this shaft count the number of revolutions of the drum, so that a five-figure reading can be obtained. Nominally, each division represents $0.1 \mu g$, and this gives a torsion head range of 1 mg. A second operating knob is geared to the drum shaft by a 1 : 4 drive, so that the complete torsion head range can be covered with this knob in 25 revolutions. The two knobs, therefore, provide a fine and coarse adjustment. A stop mechanism is built in to limit the rotation of the torsion fiber to about six revolutions and to eliminate the chance of damage due to excessive rotation.

The pans, which are made of aluminum foil, are suspended in the tubes joining the top to the base of the instrument. Access to them is by means of thermally

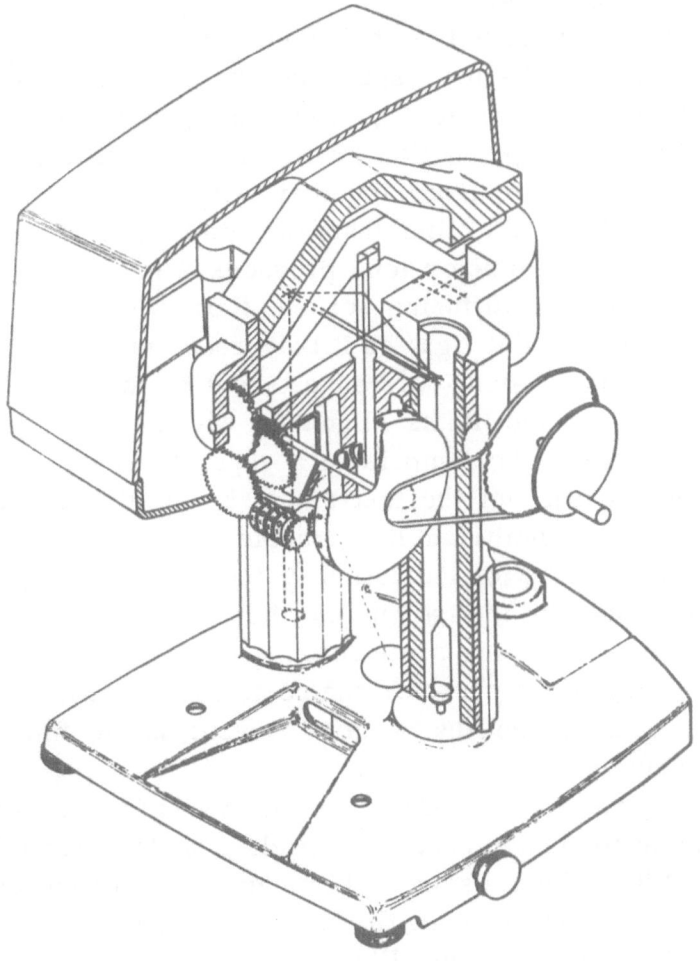

Fig. 3.

insulated sleeves which slide upward and are retained in the raised position by clips. During loading, the pans are supported on pan arrests operated by a knob on each side of the base of the balance. Rotation of these knobs to release the pans also switches on the projector lamp for the optical system.

It has already been pointed out that due to the use of the vertical beam suspension fibers, a double optical system is not necessary, and consequently a comparatively simple projection system has been fitted. A prefocus projector lamp together with a condenser lens system is used to illuminate the bottom end of the beam pointer. A projector lens produces a magnified image of this on the diffusing screen. Refraction and internal reflection of the light as it traverses the pointer result in an image having dark edges and a bright central section, which facilitates alignment with an index line on the screen. The lamp is in a separate housing spaced from the base of the instrument by an air gap, so that there is no significant heat transfer to the balance.

This form of indication of the beam position does not result in the strain on the operator produced by instruments using a microscope eyepiece. In addition, the body of the operator can be much further away from the balance, so that heating effects due to this cause are negligible.

A transport pack has been designed for the balance, so that the instrument, when packed, will withstand drops of 3 ft without damage. However, the balance can be unpacked and set up by the customer. Figure 4 shows the final instrument, and Fig. 5 a replacement beam unit.

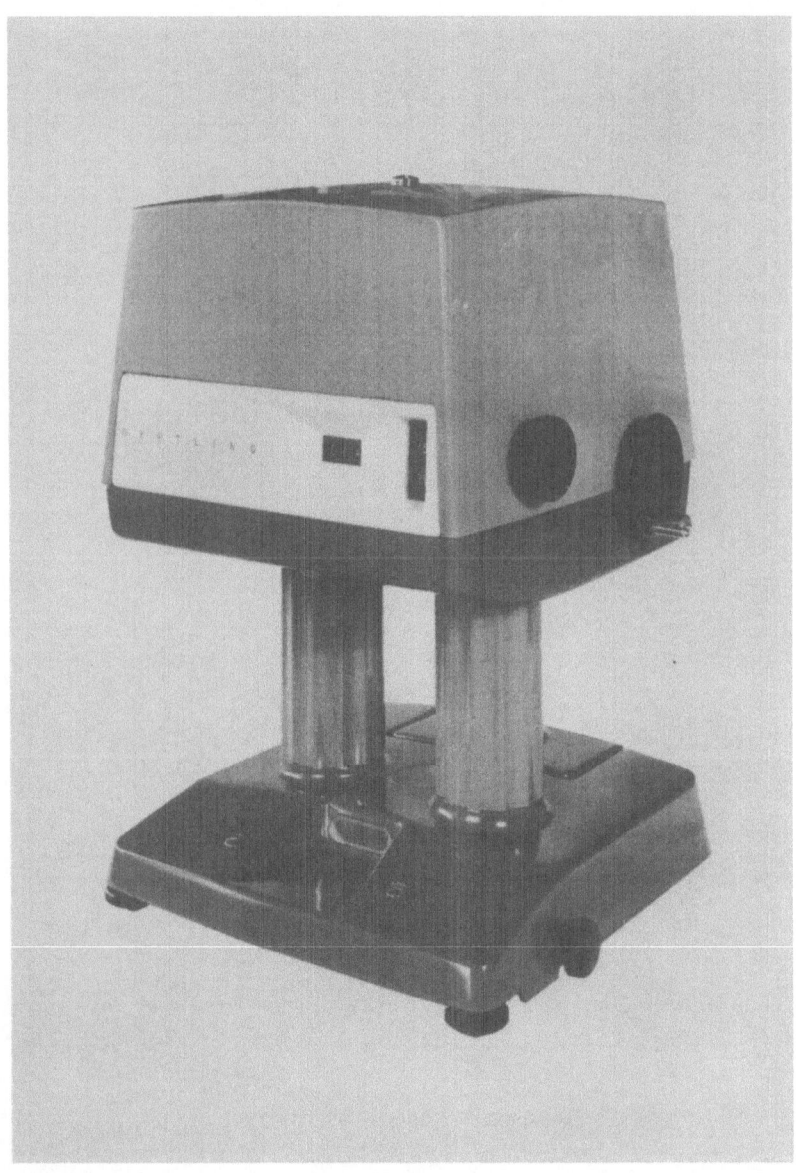

Fig. 4.

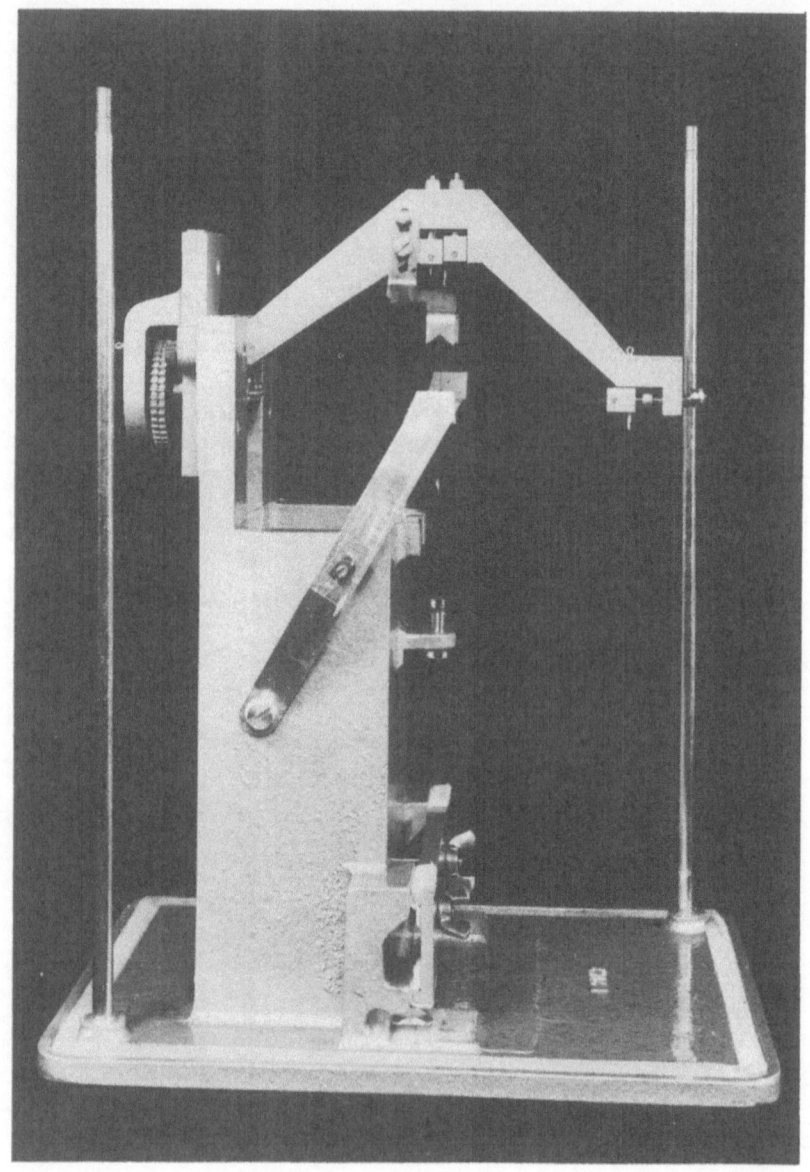

Fig. 5.

PERFORMANCE

Data on the weighing performance of a large number of beams have now been collected. Provided adequate precautions are taken, under normal balance room conditions the balance will weigh loads up to 100 mg in each pan, with a standard deviation of better than 0.08 μg. With 100 to 250 mg in each pan, a standard deviation of up to 0.12 μg can be expected. The main limitation to the reproducibility attainable is undoubtedly small dust particles settling on the pans. Several times it has been noticed that a series of consistent readings on the balance is marred by a sudden change in reading. A light brushing of the pans restores the reading to very near its original value. With cleaner conditions, a great improvement in reproducibility should be attainable. It is essential that both the balance and the forceps used to handle the load be grounded, since small electrostatic charges may be given to the pans when touched, resulting in instability of the beam.

Changes in the relative humidity of the surrounding air can cause variations in the balance zero, probably due to adsorption of water vapor on the beam and pans. The response is, however, slow, and provided that rapid changes of relative humidity are avoided, this effect does not give rise to any difficulties in use.

It has been found that the swinging system is sufficiently strong to support a load of 1 g in each pan. However, at this load the beam is considerably bent and the optical sensitivity is very low, resulting in a poor torsion head setting accuracy. In addition, at this large load, the moment of inertia of the swinging system is very much higher and the period of swing of the beam is so long that the use of the instrument becomes tedious. The friction of the air as the beam swings is sufficient to damp the beam movement quite effectively at loads up to 250 mg, but above this it becomes steadily less efficient. For these reasons, the load capacity of the instrument has been arbitrarily specified as 250 mg in each pan.

The torsion head sensitivity is determined by the length and the fourth power of the diameter of the torsion fiber. While it would be desirable to make the torsion head direct reading, so that every division represents 0.1 μg, this would necessitate the use of a torsion fiber with diameter controlled to 0.0025%, and this is clearly impossible. Practical limits on torsion head sensitivity have been set at 0.09 and 0.11 μg per division, and individual calibration for each beam is necessary. This is a simple procedure, involving the weighing of a standard weight or series of weights.

The variation of optical sensitivity with load has been found to be quite small and in general agreement with the theoretical predictions, confirming the accurate location of the vertical fibers by the transverse cylinders. However, no attempt has been made to make the arm lengths exactly equal, since the balance can be used quite satisfactorily for substitution weighing, the load being placed on the right-hand pan, and a tare weight on the left.

ACKNOWLEDGMENTS

It is a pleasure to thank Mr. J. Rock Cooper, Chairman, and the Board of Directors of L. Oertling Ltd., for permission to publish this paper. Thanks are also due to Dr. L. A. Sayce, and to Dr. G. F. Hodsman, who was formerly Technical Director of L. Oertling Ltd., for many helpful suggestions, and to Professor R. Belcher for assistance in drafting the specification of the instrument. Last, but not least, I would like to express my appreciation to Miss P. M. Ewins, who in the last five years has manufactured and tested several hundred beams.

REFERENCES

1. B. B. Cunningham, Nucleonics 5, 62 (1949).
2. H. Carmichael, Canad. J. Phys. 30, 524 (1952).
3. H. Asbury, R. Belcher, and T. S. West, Mikrochimica Acta, 598 (1956).
4. A. G. Day, Technical Report L/T256, Electrical Research Association, Greenford, Middx., England, 1951.
5. R. G. Olt, H. R. Dufour, M. I. Gray, and J. H. Wright, "A Remote Controlled Quartz Fiber Microbalance—Fabrication of Quartz Components." A.E.C. Research and Development Report MLM-1023, 1954.
6. T. T. O'Donnell, "Drawing and Working Quartz Fibers," National Academy of Sciences—National Research Council, Washington, D. C., 1958.
7. C. L. Wilson and H. M. El-Badry, Symposium on Microbalances, R. Inst. of Chemistry, London, 1950.

THE CHARACTERISTIC AND APPLICATION
OF A SIMPLE QUARTZ MICROBALANCE*

S. P. Wolsky† and E. J. Zdanuk†
Research Division, Raytheon Company
Waltham, Massachusetts

ABSTRACT

The characteristics of several quartz microbalances of the type described previously[1] have been studied. The sensitivity of these balances has been observed as a function of load. A comparison is made of the behavior of balances with 0.070-in. and 0.030-in. beams. The added utility of a balance whose sensitivity varies not too sharply with load is discussed. Further information is presented concerning the application of the vacuum microbalance to surface and sputtering investigations.

INTRODUCTION

The vacuum microbalance and associated apparatus employed in our laboratory have been described previously.[1-3] The balance is of the type using an offset beam to ensure coplanarity of the three points of support. The theory of this general type of balance and details of

*The research reported in this paper has been supported, in part, by the Electronics Research Directorate of the Air Force Research Division, ARDC, under contract AF 19(604)-8004.

†Present address: P.R. Mallory & Co., Laboratory for Physical Science, Northwest Industrial Park, Burlington, Mass.

37

its fabrication were discussed very fully at the first microbalance conference.[3-5] In this paper, we shall discuss recent efforts to increase the sensitivity of our balances (presently capable of being used to detect weight changes of $\sim 10^{-7}$ g). The need for a balance with higher sensitivity arose from our desire to investigate (1) adsorption on small-area (3–6 cm^2 total surface area) perfect single crystal samples and (2) sputtering thresholds. In the final section of this report, we shall provide further details of the application of the vacuum microbalance to the study of sputtering phenomena.

MICROBALANCE CHARACTERISTICS

The following basic relationship defines the sensitivity of a rigid-beam-type microbalance, where torsional effects may be neglected:

$$S = \frac{d\phi}{d\Delta W} = \frac{l}{Bd + a(2W_2)} \tag{1}$$

where ϕ is the turning angle in radians, l is half the length of the beam, B is the weight of the beam, d is the distance of the point of support above the center of mass, W_2 is the weight of the counterweight, W_1 is the weight of the sample, $\Delta W = W_1 - W_2$, and a is the distance from the point of suspension to the plane of the two end supports. For the ideal balance, where S is independent of load, $a = 0$.

Usually our balances are designed to provide a sensitivity of approximately $1 \cdot 10^{-7}$ g per 0.005 mm deflection at a load of about 0.5 g. It was our desire to fabricate a balance with which weight changes of approximately

$1 \cdot 10^{-8}$ g could be reproducibly detected at a load of 0.25 g. An examination of equation (1) shows that the sensitivity can be readily improved by either increasing the length or decreasing the weight of the beam. We chose to experiment with a comparatively lightweight beam. This was accomplished by reducing the beam diameter from 0.075 in. to 0.030 in. The balance was designed to provide optimum sensitivity at a load of 0.25 g. With the small-diameter beam, the effect of bending was of such magnitude as to eliminate the necessity for an offset region. In order to be able to ignore torsional resisting moments, i.e., to satisfy the condition that

$$T_W \leq 0.1 T_B \qquad (2)$$

where T_W and T_B are the torsional resisting moments of the support wires and the beam, respectively, the end and center support wires were 0.0003 in. and 0.0005 in. in diameter, respectively. The expected beam deflection was calculated using the handbook value for Young's modulus for quartz. After completion of the balance, however, we found that the actual deflection with load, as measured with a cathetometer, was approximately one half that calculated. Optimum sensitivity, therefore, was to be expected at about 0.5 g rather than at 0.25 g load.

The variation of the sensitivity of this balance, in micrograms per division, as a function of load is presented in plot A of Fig. 1. With our optical system, 1 div = 0.02 mm, and it is possible to read to $\frac{1}{4}$ div, i.e., 0.005 mm under dynamic conditions and to $\frac{1}{10}$ div, i.e., 0.002 mm when the balance is static. The sensitivity is seen to decrease at a rate of 0.36 μg/div per 0.1 g increase in load. The period of this balance was 15 sec. The balance approached the design sensitivity at a load of 0.57 g as predicted after correction for the proper bending of the beam. The sharp variation of the sensitivity with load was a result of the marked bending of the thin beam, with the consequent lowering of the center of mass.

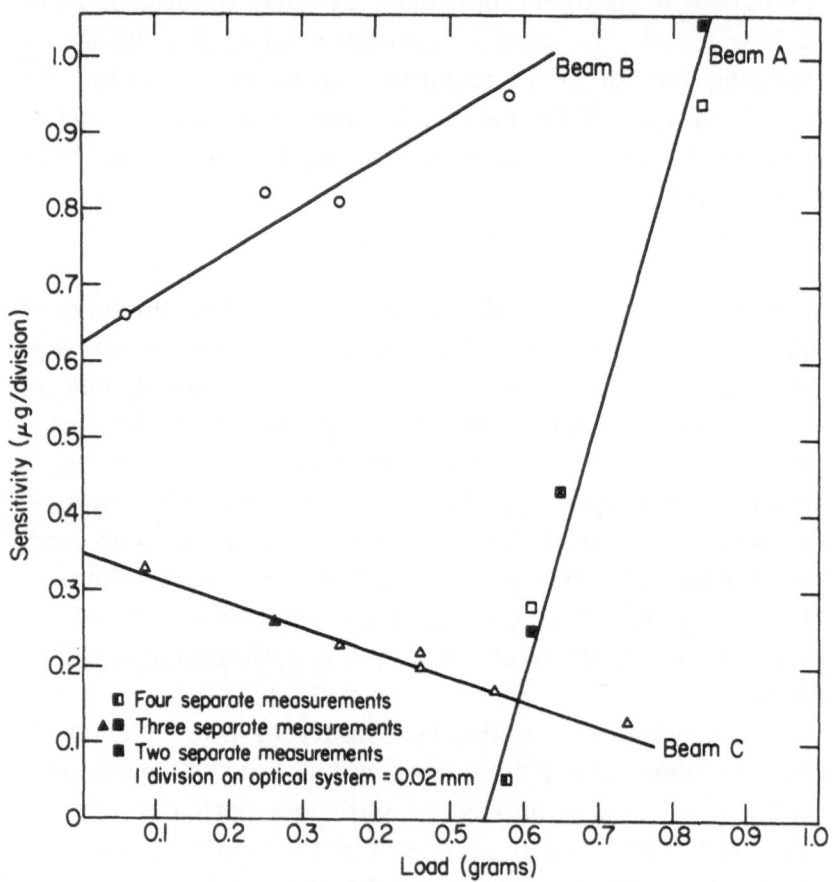

Fig. 1. Microbalance sensitivity versus load.

Below 0.57 g, the balance was unstable. The balance generally satisfied the design requirements of high sensitivity at reasonable load. Unfortunately, the application of this balance has been limited by a vibration problem. The lighter weight balance requires a special vibration-free mount that is not necessary for our conventional balances. Although this difficulty can be overcome, it is a definite disadvantage in that it introduces a complication not otherwise present.

Prior to proceeding, it is pertinent to discuss the effect upon our experiments of the variation of the sensitivity of a given balance with load. Considerable emphasis is always placed on the desirability of having a balance with a constant sensitivity over a wide range of loads. This is true for applications of the ordinary microanalytical balance, but not necessarily in our vacuum microbalance studies. It must be recognized that our experiments involve weight changes in the microgram range, and therefore, for all practical purposes, even the sensitivity of our thin-beam balance will be constant throughout an experiment at a given sample weight. Also, for certain experiments, the highest possible balance sensitivity may not be desired and would only introduce unnecessary complications. Therefore, a balance whose sensitivity can be varied simply by adjusting the load is extremely useful. In order to take advantage of this characteristic, our approach has been (1) to calibrate the balance carefully over a wide range of loads, and (2) to prepare the lightest weight samples possible so as to allow the widest latitude in the choice of the operating sensitivity as determined by the final balance loading. This may conceivably involve problems in sample preparation, but we have not found these insurmountable. The variation of the sensitivity, therefore, has not been disadvantageous, but rather has added to the versatility of the balance by allowing us to fit the sensitivity to the experimental need.

The characteristics of beam A may be compared with those of two 0.070-in. offset-beam balances recently prepared in our laboratory (Fig. 1, curves B and C). These two are especially interesting since they serve to demonstrate different types of behavior with loading. Balance B had a period of 7 sec. The sensitivity decreased at a rate of 0.060 μg/div per 0.1 g increase in load. This variation is a factor of six less than that of balance A. The decreasing sensitivity with increasing load is again the result of the lowering of the center of mass. Even when the very important effect of the bending of the beam is ignored, there exists more than one possible explanation of this type of behavior in terms of the relative position of the center and end supports. The greater rigidity of beam B has, of course, reduced the bending effect considerably in comparison to beam A.

Beam C was originally designed to have characteristics similar to those of B. In the fabrication process, however, the predetermined offsets were not obtained. The result of the fabrication error was a balance which showed increased sensitivity with increasing load, as shown in Fig. 1, curve C. The period of this balance was 13 sec. The sensitivity increased at the rate of 0.034 μg/div per 0.1 g increase in load. This variation of the sensitivity is one-half that of beam B and one-tenth that of A. The behavior of beam C resulted from the fact that the central point of suspension was near, but still below the plane of the end supports, i.e., a was negative over the range of loads investigated. In this situation two opposing effects occur as the balance is loaded, namely, (1) the raising of the center of mass because of the concentration of higher load at the end supports and (2) the lowering of the center of mass as the beam bends. If the magnitude of these effects is the same, the balance will have essentially constant sensitivity over a wide range of loads even without complete coplanarity of the points of support. The characteristics of beam C show that this ideal situation was not achieved but that effect (1) was the slightly dominant factor. It can be expected that with loads greater than those studied, the bending would become more important and the sensitivity would go through a maximum and then decrease. Sensitivities in the 10^{-8} g range were readily obtained with this balance. The increased period associated with the higher sensitivity was not objectionable in any of our experiments. We encountered no vibration problems with this balance, and its over-all behavior has been excellent.

From the above discussion, it is obvious that balance C has approached the ideal behavior for our purposes. With this 0.070-in. beam, we have high sensitivity, rigidity, and vibration-free action. The effect of load is not large enough to affect any one experiment, but is sufficient to allow a controlled variation of sensitivity according to our need. All our balances, therefore, are now being designed to achieve the characteristics of C.

THE INVESTIGATION OF SPUTTERING PHENOMENA

At the first microbalance conference we discussed briefly the application of our quartz microbalance to the investigation of sputtering phenomena from the low energy threshold region up to 5000 ev. The following presents further information on this technique. The most sensitive methods for the determination of absolute sputtering yields have a lower limit of about 10^{-4} atom/ion.[6-8] This limitation is generally imposed by the measuring instruments. Defining the minimum detectable yield, S_{min}, for the vacuum microbalance technique by the general expression

$$S_{min} = K \frac{\text{(Minimum detectable weight change)}}{\text{(Bombardment time)(Ion current)}}$$

where K is the proper constant to convert the yield to atoms/ion, it is obvious that there is no absolute sensitivity limitation. In practice the lower limit is restricted by the time necessary to sputter the minimum detectable weight change, which in most of our past work has been equivalent to 10^{14} to 10^{15} atoms. Of course, the relative rates of impinging ions and the adsorption of contaminating gases are a limitation in all sputtering experiments. In our ultrahigh-vacuum apparatus, however, the residual gas concentration is so low as to not have been a problem in any of our low-current-density work. Most important, once the zero point of the balance has been established, the process may be interrupted at any time for readings without disturbing the over-all experiment. Since the weight change is cumulative, constant energy and current density bombardments of days and weeks in duration are possible. This capability is extremely valuable for investigations of the threshold region.

Other capabilities of the apparatus are of interest in determining its usefulness in sputtering investigations.

(1) The balance sensitivity is not affected by prolonged use for sputtering experiments. In one instance, after seven months of constant use over a wide range of voltage and currents, the microbalance showed no significant change of its sensitivity. (2) Adsorption measurements allow the monitoring of surface area changes as a result of ion bombardment. The surface area of the sample can have a marked effect upon the observed sputtering yield.

The vacuum microbalance technique provides an excellent means of investigating the sputtering of adsorbed gases and surface films in a controlled manner. Very little is known on this important subject. The type of information which may be obtained is probably best illustrated from a step by step consideration of the following possible experiments performed completely *in situ*: (1) sputtering to a clean, reproducible surface condition as evidenced by the measured yield, (2) exposure to a given ambient as a function of time and temperature with well-defined adsorption or film formation being followed through the observed weight change, (3) sputtering of the sample at low current density as a function of time until the clean surface yield is obtained. Surface area measurements can, of course, be taken at any desired point in the experimental sequence. Studies can be made with surface films ranging in thickness from monolayers to angstroms. A comparison of yields for films prepared under different conditions should also be extremely interesting. Considerable information on film structure may be obtained from the sputtering of well-defined surface films. The microbalance technique, therefore, appears to offer a unique opportunity to employ sputtering as a surface study tool.

ACKNOWLEDGMENT

The authors wish to acknowledge the assistance of R. Chase and R. Leighton in the fabrication of the balances and of Mrs. P. Rodriguez for her help in the development of the experimental techniques.

REFERENCES

1. S. P. Wolsky, Phys. Rev. 108, 1131 (1957).
2. S. P. Wolsky, Semiconductor Products, June, 1959.
3. S. P. Wolsky and E. J. Zdanuk, Vacuum Microbalance Techniques, Vol. 1, Plenum Press, New York, 1961, p. 35.
4. E. A. Gulbransen and K. F. Andrew, Vacuum Microbalance Techniques, Vol. 1, Plenum Press, New York, 1961, p. 1.
5. R. F. Walker, Vacuum Microbalance Techniques, Vol. 1, Plenum Press, New York, 1961, p. 87.
6. R. V. Stuart and G. K. Wehner, Phys. Rev. Letters 4, 409 (1960).
7. N. D. Morgulis and V. J. Tischenko, Soviet Physics JETP 3, 52 (1956).
8. D. McKeown, Rev. Scient. Inst. 32, 133 (1961).

TUNGSTEN HELICAL-SPRING MICROBALANCE

Samuel L. Madorsky
National Bureau of Standards
Washington, D. C.

ABSTRACT

This paper describes a helical-spring microbalance made of 3-mil tungsten wire and enclosed in a Pyrex glass housing. It was used in connection with a problem involving measurement of rates of thermal degradation of polymers in a vacuum by the loss-of-weight method. The balance has a sensitivity of about 550 μ/mg and is so constructed as to avoid errors that might be caused by the thermal expansion of its parts or of the supports holding the balance and the cathetometer used in connection with it.

INTRODUCTION

A tungsten helical-spring microbalance was developed at the National Bureau of Standards in connection with a problem involving a study of rates of thermal degradation of polymers. The study consisted of measuring the rates of weight loss of 5-mg samples of the polymers during destructive distillation in a vacuum or in helium at about 200 to 520° C.[1] The weight of the crucible holding the samples was about 200 mg, and that of the wire from

47

which the crucible was suspended was about 100 mg. A diagrammatic view of the balance is shown in Fig. 1. In this figure, part I-a fits on top of part I-b. The spring is shown enclosed in a Pyrex housing.

Quartz springs had previously been used in connection with vacuum weighing, for example, in the McBain-Bakr sorption balance.[2] However, quartz springs are very fragile; springs of tungsten are much sturdier. Tungsten was selected in preference to other metals as a material for the spring because it has almost the same coefficient of thermal expansion as Pyrex glass. The linear thermal expansion of tungsten is $4.2 \cdot 10^{-6}/°C$ and that of Pyrex is $3.6 \cdot 10^{-6}/°C$.[3] Another advantage of tungsten over quartz lies in the fact that the low internal friction of quartz produces a very small damping action, and when the spring is disturbed, particularly when in a high vacuum, several hours may be required for the spring to come to rest. Tungsten has much higher damping, and it takes only 15–20 min for the spring in the balance to come to rest.

The following were the characteristic features of the helical-spring balance:

Diameter of tungsten wire . . . . 3 mils
Diameter of coils . . . . 1.1 cm
Number of coils 23
Total elongation under 0.3-g weight . 16 cm
                                (approximately)

In constructing the spring, a 100-g weight was suspended from one end of a tungsten wire about 1 m long. The other end of the wire was attached to one end of a threaded steel rod held firmly in horizontal position in

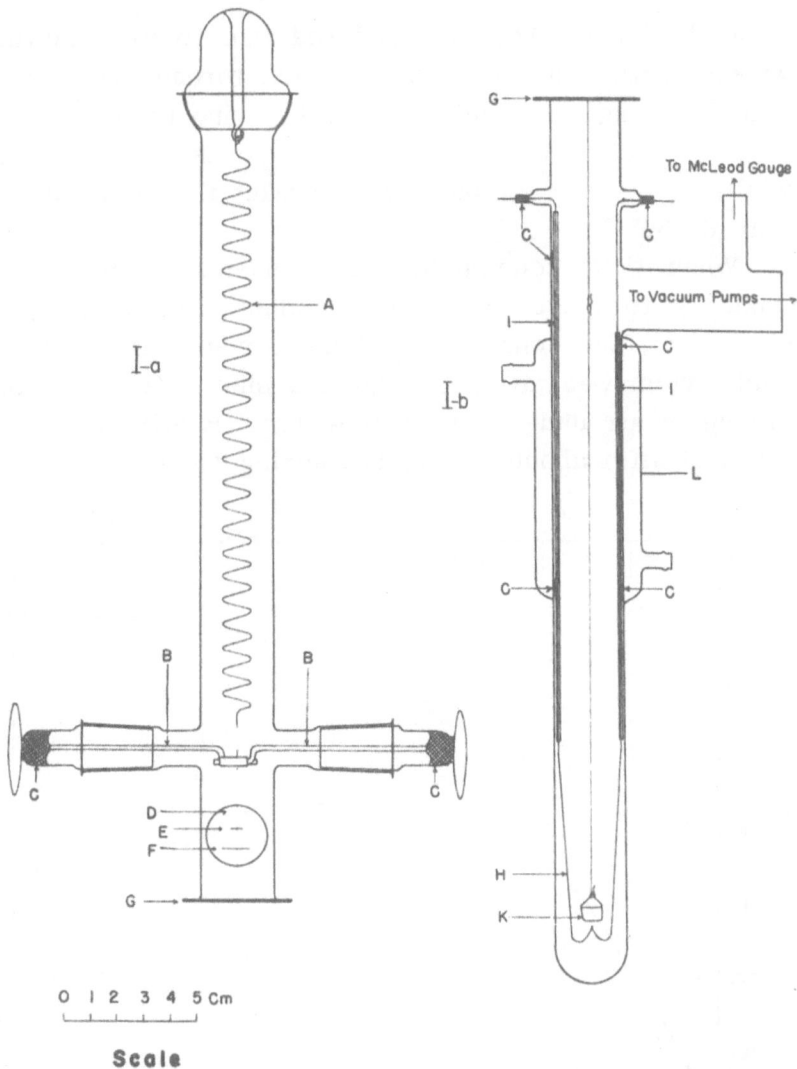

Fig. 1. Tungsten spring balance. Part I-a fits on top of part I-b. A, tungsten spring; B, brass friction clasps padded with rubber sleeves; C, hard apiezon wax; D, flat Pyrex window; E, crossline on spring extension; F, crossline on window; G, ground flanges; H, chromel-constantan thermocouple; I, glass capillaries for holding the thermocouple wires inside the apparatus in fixed positions; K, platinum crucible; L, water jacket.

a steel sleeve. The threaded rod was rotated and the wire directed into the groove of the thread. The completed spring was held in place by tying the loose end of the wire to the rod. The spring was then annealed while it was still on the rod by heating it in an oven for 6 hr at 250° C.

When first constructed, the tungsten spring had a tendency to increase in length when loaded (delayed elasticity). This tendency continued for some time before equilibrium was attained. Figure 2 shows creep of the spring in air under a load of 400 mg. Equilibrium was attained after about 160 hr. The spring was then allowed

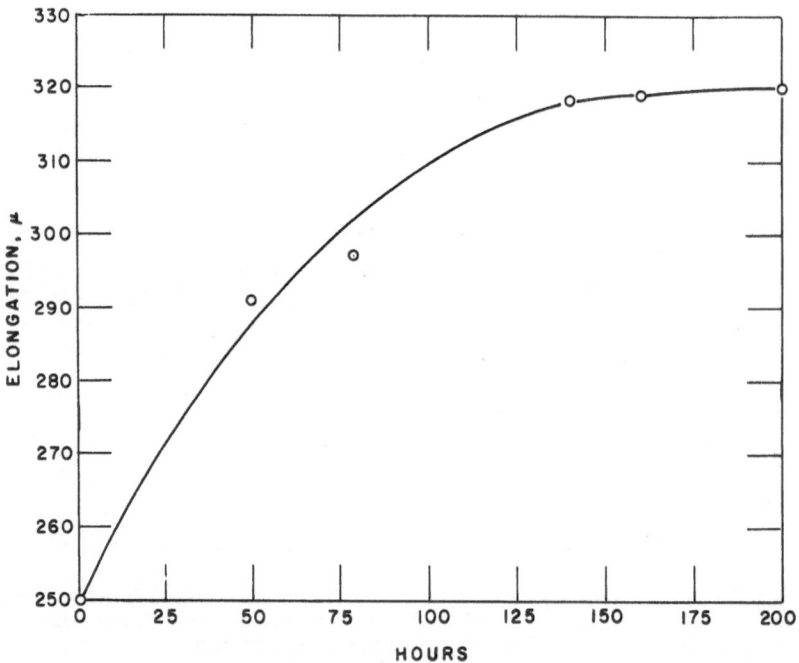

Fig. 2. Creep of the newly constructed tungsten helical spring.

to hang under the same load for a total of 30 days before it was used in the apparatus. After this no further deformation was detected.

In order to minimize the effect of further deformation when the spring is not in actual use and to avoid disturbance of the spring while the crucible is being suspended or removed from it, the balance is provided with a pair of friction clasps BB, as shown in Fig. 1. The clasps are operated by rotating them in opposite directions until they grip the wire extending from the spring. By lifting part I-a and the crucible above part I-b (Fig. 1), the crucible becomes accessible.

Readings of the position of the crossline E on the spring extension are made with respect to line F on the window D by means of a sensitive cathetometer. By making line F serve as a reference for measurement, errors due to the thermal expansion of the balance supports and cathetometer stand are eliminated. Part I-b of the apparatus is provided with a jacket through which water at room temperature is circulated to keep the spring cool.

The spring balance described has been in operation for about 10 years. Its sensitivity, λ, was 530 μ/mg.* This sensitivity has gradually increased over the 10-year period, as shown in Table 1. However, each experiment involves a check of the sensitivity, since the crucible and sample are weighed before and after each experiment on a semimicrobalance, which has a sensitivity of

*This value is taken from Table 2 in reference [1]. At the end of the first 7 experiments shown in this table the spring inadvertently received a violent jolt, which caused an increase in the average value of λ during the next 15 experiments from 525 to 530 μ/mg.

Samuel L. Madorsky

TABLE 1

Change of Sensitivity, λ, of the
Tungsten Spring Balance

Year		λ, μ/mg
1952	average	530
1956	average	546
1961	Expt.	
	1	575
	2	578
	3	582
	4	581
	5	581
	6	581
	7	585
	8	582
	9	580
	10	580
	11	576
Average for 1961		580

0.005 mg. Some data on the average sensitivity are given in Table 2, which shows the results of a series of experiments carried out in 1952.[1]

The performance of the tungsten spring balance can best be illustrated by describing a study of rates of thermal degradation of polytetrafluoroethylene.[4,5] Each experiment consisted in heating a sample in a vacuum at some constant temperature and observing the rates

TABLE 2

Average Sensitivity, λ, of the Tungsten Spring Balance

Temperature, °C	Duration, hr	Weight of sample, mg	Total loss in weight, mg	Sensitivity of spring, μ/mg*
335	450	5.81	3.55	526
340	425	5.54	2.94	526
345	375	5.47	3.68	534
350	360	5.45	4.40	533
355	310	5.13	4.56	530
385	510	6.55	3.24	524
390	450	6.47	3.80	521
395	430	6.46	4.54	526
400	270	6.07	4.69	525
403	270	6.03	5.16	528
335	463	4.62	2.31	532
340	420	5.29	3.25	532
345	420	5.25	3.78	532
350	360	5.45	4.40	533
355	272	5.54	4.58	538

*Average sensitivity λ = 530 μ/mg.

of loss of weight of the sample. Two series of experiments were carried out. In one series at higher temperatures,[4] the initial rates were as high as about 0.1% to 1.0% of the sample per minute, as shown in Fig. 3. In another series of experiments, at lower temperatures,[5] the initial rates were low, varying from about 0.02 to

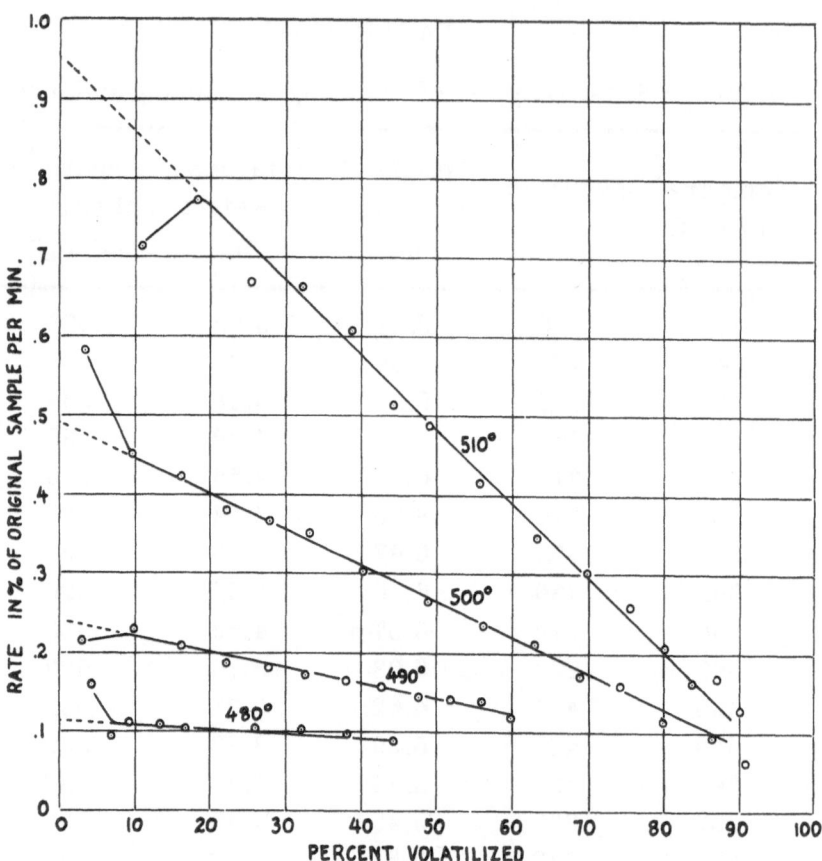

Fig. 3. Rates of thermal degradation of polytetrafluoroethylene at higher temperatures.

0.17%/min, as shown in Fig. 4. By plotting the logarithms of the initial rates for the given temperatures as a function of the inverse of the corresponding absolute temperatures, a straight line (Fig. 5) is obtained, whose slope represents the activation energy of the reaction or reactions involved in the degradation of the poly-

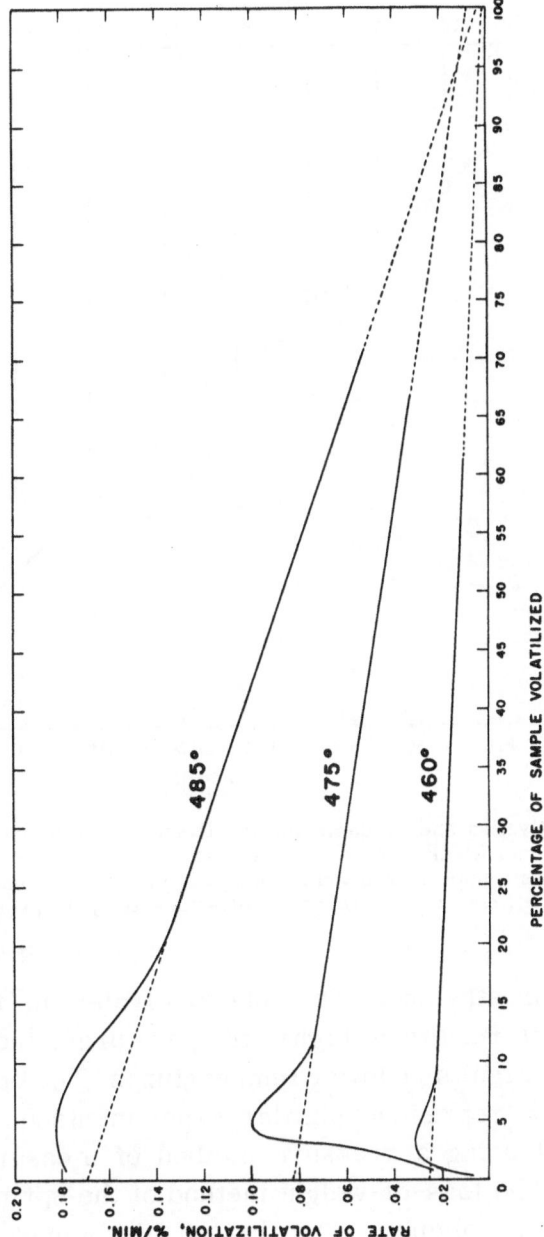

Fig. 4. Rates of thermal degradation of polytetrafluoroethylene at lower temperatures.

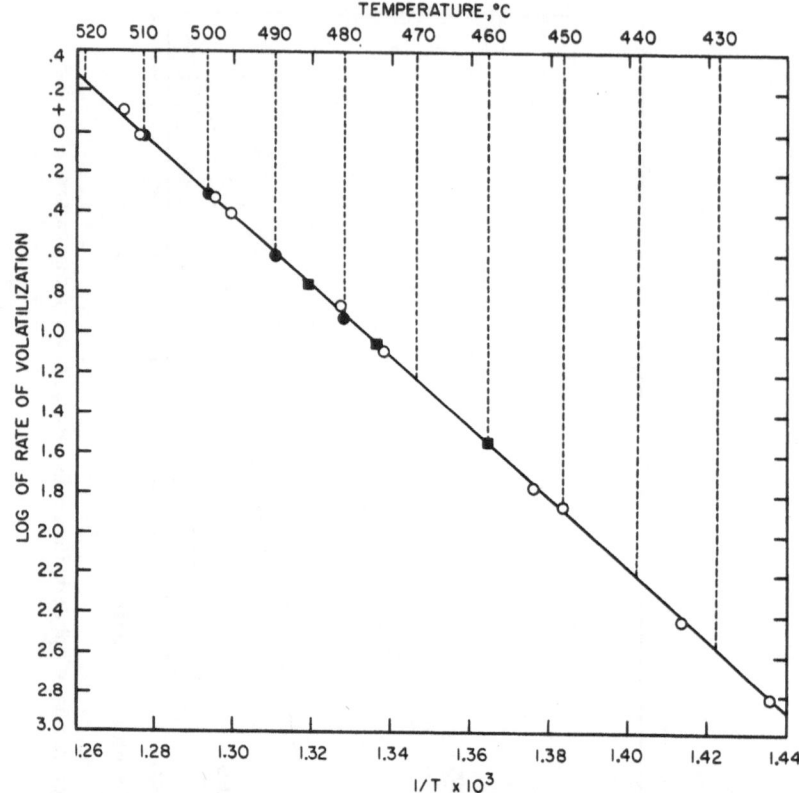

Fig. 5. Activation energy curve for the thermal degradation of polytetra-fluoroethylene. Black circles represent experiments in the spring balance at the higher temperatures and black squares represent those at the lower temperatures; open circles represent experiments by the pressure method.

tetrafluoroethylene. The black circles in the figure represent results at higher temperatures, and the black squares results at lower temperatures. The open circles in Fig. 5 represent similar experiments on the same material using a pressure method of measurement instead of the loss-of-weight method of the spring balance. Since this polymer, when heated, decomposes into the monomer, the rate of degradation can be followed by measuring the change of pressure in the apparatus.

This illustration shows that there is excellent agreement of results in these two completely different methods of measurement and that there is good reproducibility of performance in the tungsten helical-spring balance.

REFERENCES

1. S. L. Madorsky, J. Polymer Sci. 9, 133 (1952).
2. J. W. McBain and A. M. Bakr, J. Am. Chem. Soc. 48, 690 (1926).
3. Handbook of Chemistry and Physics, Edition of 1960, The Chemical Rubber Publishing Company, Cleveland, Ohio.
4. S. L. Madorsky, V. E. Hart, S. Straus, and V. A. Sedlak, J. Research Natl. Bur. Standards 51, 327 (1953).
5. S. L. Madorsky and S. Straus, J. Research Natl. Bur. Standards 64A, 513 (1960).

VACUUM SYSTEM FOR USE WITH A MICROBALANCE*

Glenn F. Rouse
National Bureau of Standards

A clean system and a very high vacuum are required for many experimental studies in which a microbalance is used, especially when surface phenomena are under investigation.

Since a great amount of information on general design features and operating practices for ultrahigh-vacuum systems is available, one would not anticipate much difficulty in the design of a system for use with a microbalance. However, the incorporation of special features which assure convenience and safety in handling, mounting, and using the microbalance is not a simple matter. It does, in fact, pose many problems.

The system which will be discussed satisfies the conditions of cleanliness and low pressure. Considerable operating experience indicates that it has the desired features of convenience and safety.

GENERAL FEATURES OF THE SYSTEM

A photograph of the system is shown in Fig. 1. With the exception of a few glass parts it is made entirely of metal, most of which is type 304 stainless steel. A small amount of Kovar is needed to allow for glass sealing, a monel bellows is used where a means of adjustment is

*This project was supported in part by the Air Force Cambridge Research Center.

Glenn F. Rouse

Fig. 1. Vacuum system for use with a microbalance.

required, a lightweight aluminum dial is mounted on the balance shaft, and a small amount of inconel is used because of its low value of magnetic permeability.

All permanent joints are made with high-melting-point solder or by heliarc or electron beam welding. All vacuum seals, other than that of the box to its base plate, are made by means of OFHC copper gaskets (thickness approximately 0.025 in.) clamped between opposing knife edges.[1] Any other proven form of metal gasket seal could be used.

The rectangular enclosure N (11.25 in. by 5 in. by 3.5 in.) which houses the microbalance consists of a stainless steel base plate (thickness 0.5 in.) and a cover which was made by heliarc welding five suitably shaped pieces of type 304 stainless steel sheet (thickness 0.125 in.). A gasket of aluminum foil (thickness 0.005 in.) clamped between a flat polished area on the base plate and the polished edge of the cover produces a vacuum-tight seal (see Fig. 2).[2]

Reentrant tubes (not visible), with windows sealed to their ends, accommodate the optical unit O. This unit consists of two microscopes and a comparison eyepiece. It established the zero or reference position for the balance beam and serves to detect any angular displacement of the beam from this position.[3] Some details of the optical system are shown schematically in Fig. 3. A comparison eyepiece having the desired width could not be obtained. Total reflecting prisms serve to accommodate the width of the eyepiece to that required for the balance.

The part labeled P encloses the sample whose weight

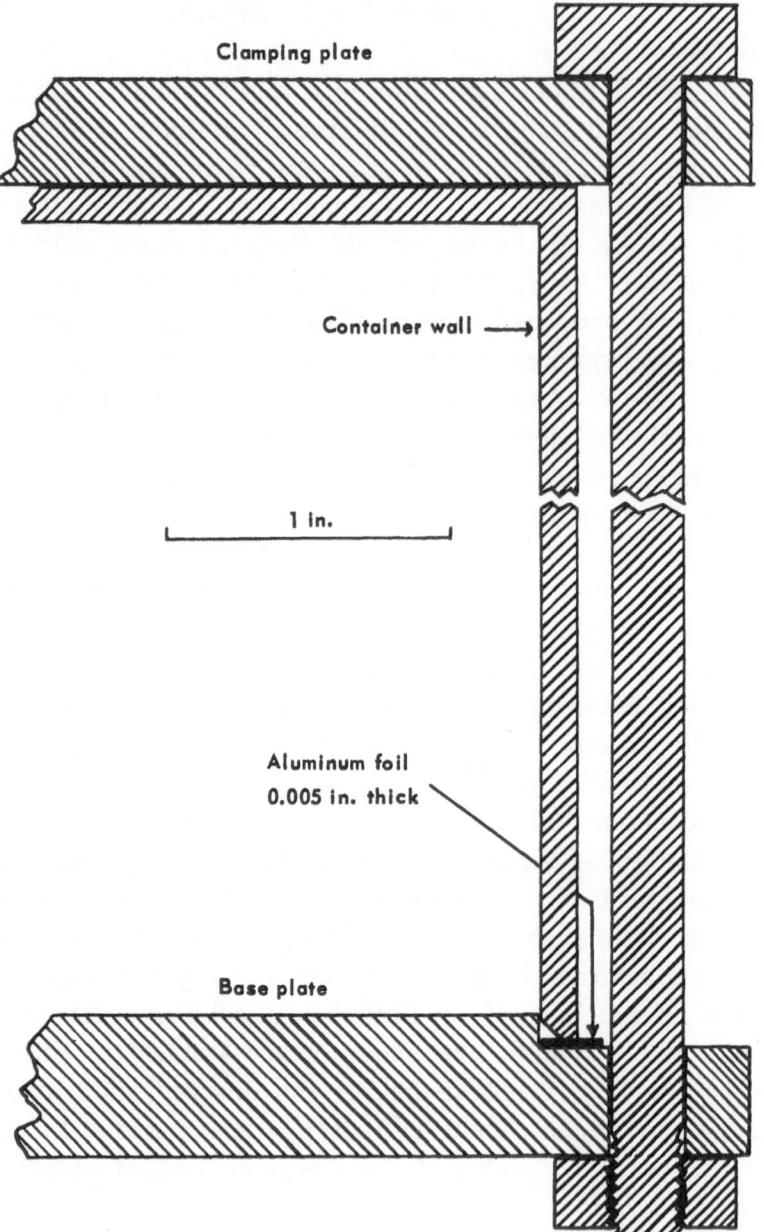

Fig. 2. Section of microbalance container.

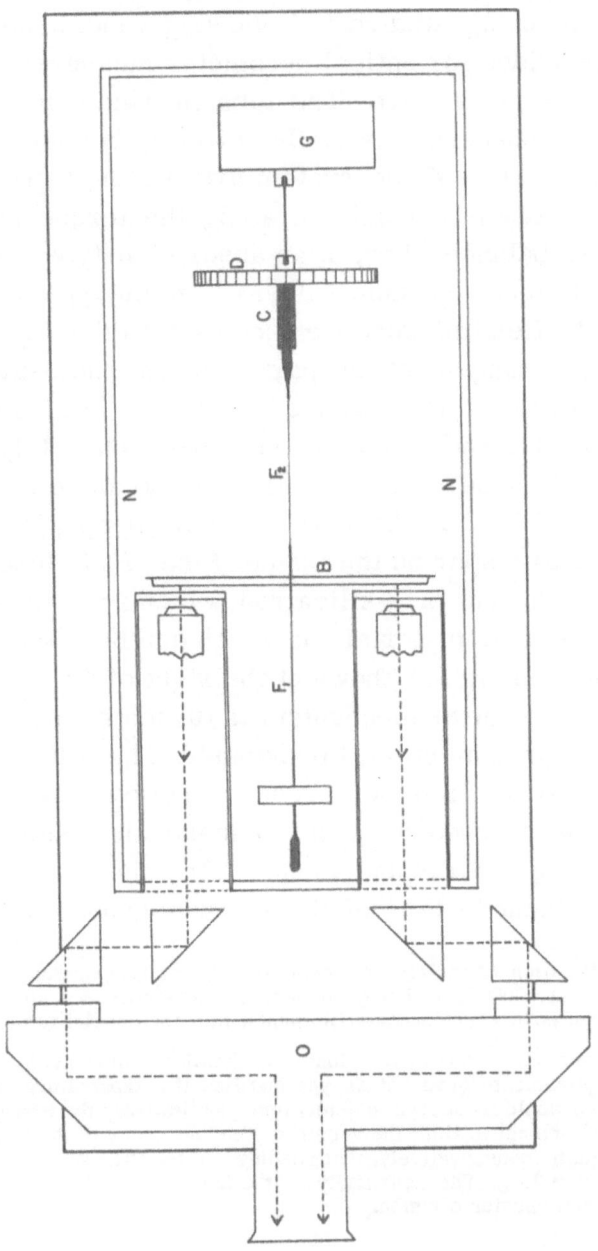

Fig. 3. Schematic of optical system.

change is being studied.* A viewing window and a tube which provides for optical pyrometer measurements are attached to this part. The tube in which the counter-balance is suspended is visible directly behind P.

Other parts of the system are: R, an electromagnet which makes it possible to apply the torque needed to twist the balance fiber; S, an absorption-type forepump; T, an all-metal vacuum valve; U, an ion-type pump; and V, a wide-field microscope used to read the torsion dial.

The arrangement of parts within enclosure N is shown in Fig. 4. The balance† (B, F_1, F_2), which is made entirely of fused silica, is supported on a tripod-type frame (A) made of fused silica rod (diameter approximately 0.125 in.). The bow to which fiber F_1 is attached is fused to a spur on the frame. Fiber F_2 is fused to the drawn-down end of a silica rod, which fits into a sleeve (C) of the same material and which can be rotated by the gear mechanism (G) shown at the right of the figure. The worm in the latter mechanism is turned by means of the electromagnet mentioned previously. The beam rests are raised and lowered by screw and bellows mechanisms. The scale marked on drum D makes it possible to determine angular displacements of the shaft to which fiber F_2 is attached. One of the balance hangdowns passes

---

*The sublimation of metals and metal alloys, in a temperature range centering around 1000°C, is being studied. The specimen is a small cylinder which is heated by a pure tungsten filament within and coaxial to the cylinder.

†The balance was made by the Monsanto Chemical Company, Miamesburg, Ohio. Requirements have not as yet justified the expenditure of time and effort which would be needed to determine quantitatively the extent to which certain factors act to limit the sensitivity and accuracy of the balance. One can say, quite conservatively, that readings on the balance are accurate to the nearest 0.2 μg. The sensitivity of the balance is 0.05 μg per readable torsion-drum vernier division.

Fig. 4. Detail inside microbalance container.

through port E, the other through a similar port at the rear of the baseplate.

COMMENTS

Several useful practical hints are presented in the following comments.

1. Aluminum companies can supply large-size sheets of foil whose surfaces are remarkably free of imperfections or scratches. A satisfactory gasket of any size and shape can be cut out provided the procedure used guards against damage to the surfaces. The gasket used in this system is obtained by clamping the foil between two metal frames having the required size and shape. The surfaces which contact the foil are polished, and pinning prevents slippage of one frame with respect to the other. Excess foil is trimmed off with a sharp knife.

2. There is a tendency for aluminum foil to stick to the surfaces of the objects between which it is clamped.[2] The cover should be designed so that it is possible to apply the force required to free it from the base plate.

3. Since the coefficient of expansion of type 304 stainless steel is approximately $19 \cdot 10^{-6}$ per degree centigrade and that of fused silica is very small, care must be taken to avoid setting up dangerous stresses during bakeout. Note that in this system one leg of the fused silica frame rests in a hole, another in a groove, and the third on a flat. The latter two legs can slide quite freely on their supports. The metal strip which connects the gear mechanism to the silica frame is fitted loosely so as to permit relative motion.

4. In experiments in which the surface condition of the test specimen is important, the procedure followed in loading the balance should be so planned as not to alter this surface condition. In an all-glass system the loading procedure involves a glass blowing operation which might be highly undesirable because of the high temperature and because of possible contamination from the flame.

For the system described here loading does not involve a change in temperature. The port through which the specimen is inserted can be covered as soon as the operation is finished. By use of a torque wrench and screw driver the joint can be made vacuum-tight in about ten minutes.

5. The total change in mass which can be measured by twisting fiber F_2 is at most approximately 400 μg. There is an alternative method of balancing which is capable of a much greater range.

If the counterweight, or a portion of it, is a small bar magnet (a piece of Cunife wire,[4] diameter 0.025 in., length 2 cm, has been used) the direct current in a single-layer solenoid surrounding the magnet can be adjusted to balance the beam. It is of course necessary that the tube in which the counterbalance hangs be nonmagnetic. In this system, the tube is inconel.

The force which the field exerts on the magnet can be expressed as $F = kI$. The value of the constant k is dependent upon various factors, an important one being the position of the magnet relative to the center of the solenoid. This latter fact makes possible a convenient control over the value of k but it also necessitates the exercise of considerable care to avoid unwanted changes in k.

The following data are illustrative of what can be obtained by this method of balancing. Calibrated weights were used singly and in various combinations to vary F from 100 to 1500 μg and corresponding values of I were measured. The value of k was calculated to be 3.506 ± 0.006 μg/ma.

6. That part of the system designated P (Fig. 1) may be replaced by another of different design. The only requirement is that the new structure be provided with a flange which matches the one on the system. For example, if one wishes to heat a test specimen inductively the substitute structure would be a tube, all or part of which is made of suitable nonconducting material.

7. In some microbalances the fibers which suspend the beam are cemented to the supporting frame. Silver chloride (melting point 450° C) is sometimes used as a cement. The advisability of using cement in a system which is to be baked is open to question partly because of the vapor pressure of the cement and partly because of the mechanical nonlinearity and creep of such joints. Note that in this system the use of cement is avoided by fusing silica to silica.

8. An attempt has not been made as yet to measure the ultimate vacuum that can be attained in this system. The VacIon current becomes so small as to be undetectable on the power supply meter (82 mm scale—20 μa). This indicates a pressure of less than 10^{-8} mm Hg.

REFERENCES

1. G. W. Hees, W. Eaton, and J. Lech, Vacuum 4, No. 4, 438 (1954).
2. S. Ruthberg and J. E. Creedon, Rev. Sci. Instr. 26, 1208 (1955).
3. Hugh Carmichael, Canadian J. Phys. 30, 524 (1952).
4. I. L. Cooter and R. E. Mundy, J. Research Natl. Bur. Standards 59, 379 (1957).

MASS AND THERMAL MEASUREMENT WITH RESONATING CRYSTALLINE QUARTZ

A. W. Warner and C. D. Stockbridge
Bell Telephone Laboratories, Inc.
Whippany, New Jersey

ABSTRACT

Certain quartz crystal plates are frequency sensitive to mass loading when operated in particular modes of vibration. Frequency can currently be measured with a precision as high as one part in 10^{10} in as short a time as one second so that dynamic measurement of extremely small mass changes can be made. Further, the resonant frequency is a function of temperature gradients within the quartz plate so that certain plates can be used to detect heat effects from monolayer sorption phenomena on less than 1 cm^2 of surface.

The importance of crystal plate design and its related temperature control is discussed; graphs to aid the choice of a correct design are given.

The relevance of suitable oscillator circuitry is also discussed, since the circuit must be relied upon both to define or delineate the resonant frequency of the quartz plate and to maintain a known amplitude of vibration. Circuitry for use at 5 Mc is given.

With edge-mounted quartz plates vibrating in the

thickness shear mode, detection of mass changes in the range of 1 to 10 picograms seems feasible.

INTRODUCTION

The use of quartz plates to detect changes in mass is not new. The effect of added mass on crystal frequency has been known since the early days of radio, when frequency adjustment was accomplished by a pencil mark on the controlling quartz crystal. A recent detailed study of the effect has been carried out by Sauerbrey.[1] In the present paper experience in the design and fabrication of ultraprecise quartz crystal units for frequency control[2] is used to demonstrate more fully the design parameters affecting the sensitivity and accuracy of this method of mass measurement.

The proper terminology for this system of mass measurement is not yet established, but we believe that the generic term, micro-weighing, applies. The presence of mass, when tightly bound to a quartz plate, is detected by measuring its moment of inertia at high frequency. Calibration can be effected using the known characteristics of crystalline quartz or by observing the effect on the resonant frequency of a known mass evenly* distributed over a given area of the quartz plate.

The field of usefulness is limited to those materials which can be uniformly bonded to the quartz plate and which will not unduly lower the Q of the resonator. Within this field, however, extremely small mass changes can be readily detected and recorded at intervals as small as one second.

*This requirement is stringent.

The best quartz plate design for a given micro-weighing experiment is one with the highest significant sensitivity to mass. That is, the ratio of the frequency change with mass with respect to any other source of frequency change must be high. Thus the ability to define and measure the resonant frequency is often more important than the intrinsic frequency sensitivity to mass of the crystal plate. Other factors influencing the frequency stability include the effects of temperature variation, amplitude of vibration, Q, electrical impedance, and mechanical support.

The main purpose of this paper will be to discuss (1) the effect of temperature variations on the resonant frequency and (2) the effect of oscillating and measuring equipment on the frequency measured and recorded. Consideration of these effects leads to the proper choice of quartz plate design for any given set of experimental conditions.

For convenience in design, a term is required to express the minimum detectable mass change which can be significantly measured. The frequency uncertainty caused by factors other than mass change may be plotted in terms of the equivalent mass uncertainty. For brevity we term the resulting frequency uncertainty, expressed in picograms, the "Effective Mass Sensitivity," or E.M.S.

This term, E.M.S., applies when all factors affecting a measured frequency are simultaneously operating. However, for the sake of clarity in developing the illustrations we have made use of the term while considering each factor separately, thereby making the assumption that the factors not considered are insignificant. Where more than one factor is significant, as is particularly the case for an optimum design, some method is needed to calculate the resultant E.M.S. With a knowledge of the behavior of one or more variables, the minimum detectable mass change could well be less than the root mean square addition appropriate when the variables are independent and random. A precise method of arriving at the true E.M.S. will, therefore, be postponed until more experience has been obtained.

DESIGN CONSIDERATIONS—QUARTZ PLATES

The most suitable quartz resonator for micro-weighing is the AT cut, vibrating in thickness shear. Both the ease of selection (by crystallographic cut) and the magnitude of the temperature coefficient of frequency are superior to those of other modes of vibration. The temperature at which the zero temperature coefficient occurs can be selected by precise control of the angle at which the quartz plate is cut with respect to its crystallographic axes, as seen in Fig. 1. Figure 2 shows a typical family of frequency vs temperature curves for the AT cut. Each curve represents a slightly different angle of rotation. It can be seen that in the vicinity of 27° C the temperature-frequency curves can be nearly flat. Furthermore, AT-cut plates can be designed to have a simple resonance involving only quartz and the electrode material. The resonance is free of coupling to other modes of vibration and quite isolated from the edge of the plate and its support structure.

The lower frequency cuts, CT, DT, GT,[3] etc., all suffer from mounting loss, since the resonance cannot be decoupled from the support except by use of a reso-

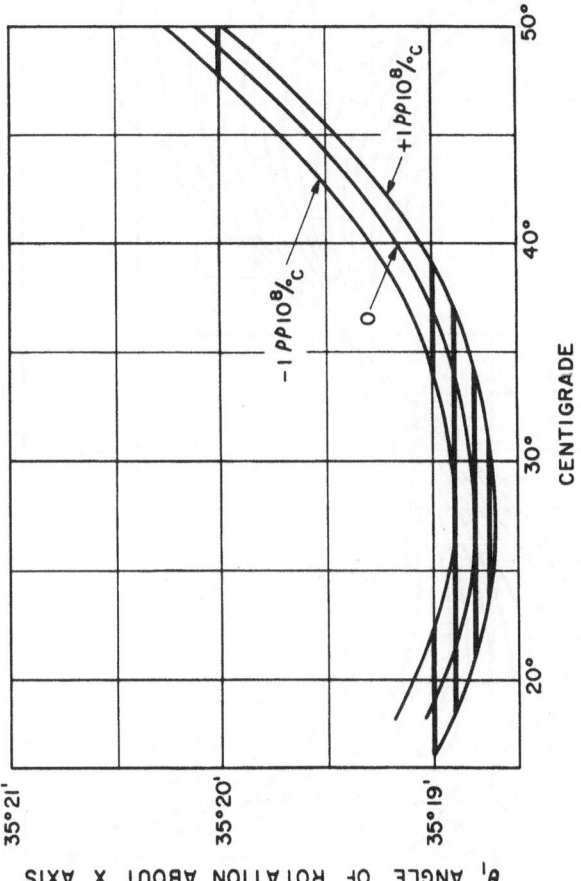

Fig. 1. Relationship between the angle of cut and the temperature of zero temperature coefficient. Also shown are limit curves for 1 part in 10^8 per degree centigrade.

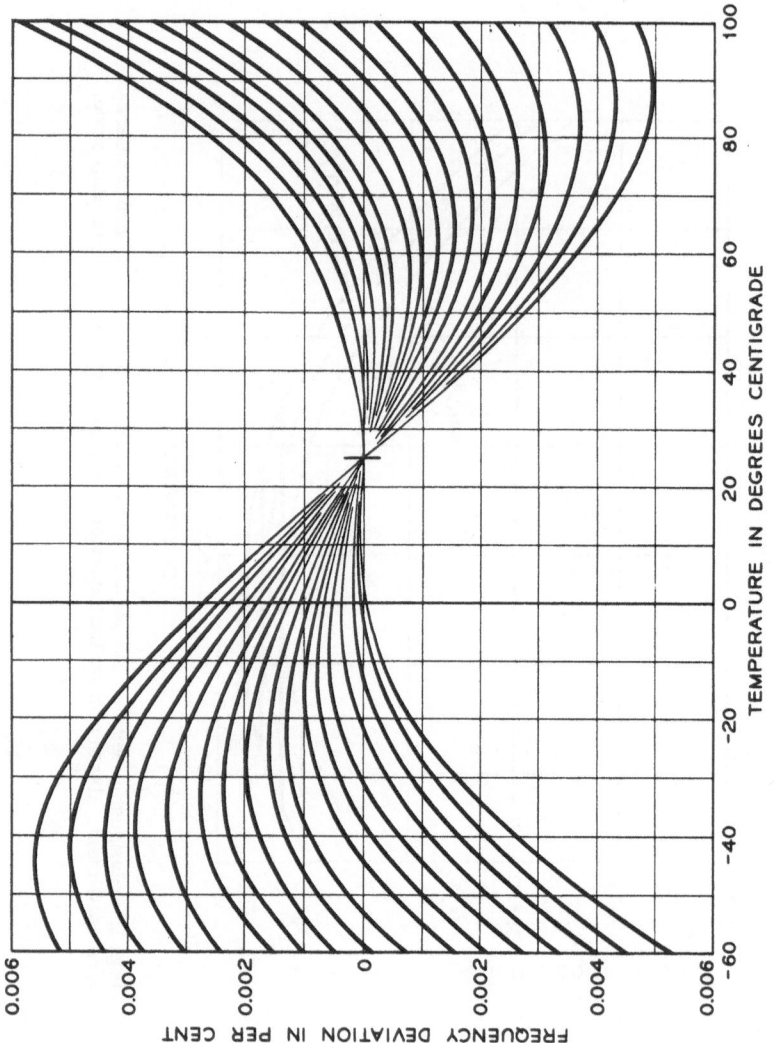

Fig. 2. Frequency vs. temperature for different zero-coefficient temperatures.

nant mounting. The result is that the mounting structure becomes part of the mechanically vibrating, frequency-determining system, with a deleterious sensitivity to external shock, lowered Q, and frequency drift with time.

CONTROL OF TEMPERATURE

Figure 3 shows the effect of temperature and frequency on E.M.S. for AT-cut plates. Three curves are shown, representing three different abilities of apparatus to control ambient temperature. In constructing these curves, the temperature coefficient has been taken as one part in 10^8 per degree centigrade, a value which is not exceeded over a range of about $\pm 2°$ C from a zero temperature coefficient point near room temperature. Figure 1 shows that this value is reasonable as a design criterion. In plotting Fig. 3, it has been assumed there are no significant frequency-determining factors other than temperature operating. That this will not be the case is implied by breaking the lines in areas of lowest E.M.S. The slope of the lines is a result of the equation[3] $f = k/t$, where $k = 1700$ kc-mm, f is the frequency of resonance, and t the thickness of the AT-cut quartz plate. By differentiating, $df/f = (-f/k)\, dt$; hence, the percentage change in frequency with a change in thickness (or mass) is directly proportional to the frequency. The corresponding thickness change has been converted to picograms (10^{-12} g), taking the density of quartz as 2.65 g/cm^3. Experimentally, this has been verified to within a few percent by observing the effect of added known masses of gold and aluminum.

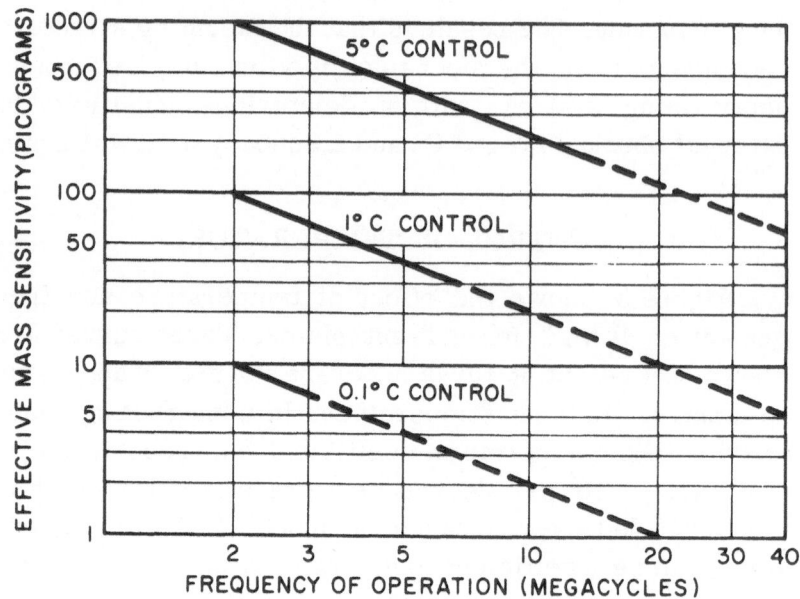

Fig. 3. Limiting effect of temperature on E.M.S. vs. frequency. Area of contact 10 mm².

CONTROL OF AMPLITUDE OF VIBRATION

One of the limitations in micro-weighing by this method, implied by the broken curves in Fig. 3, is that the frequency is to some extent a function of amplitude of vibration of the quartz plate. At low amplitudes the effect is small, but there are limits to small signal amplification. At the lowest practical quartz plate driving levels, in the microampere range, and with 1 db control over these levels, the limitation is approximately as shown by the heavy line in Fig. 4. The reasons for this behavior are not clearly understood. The effect is not due to dissipation, since the total power input to the

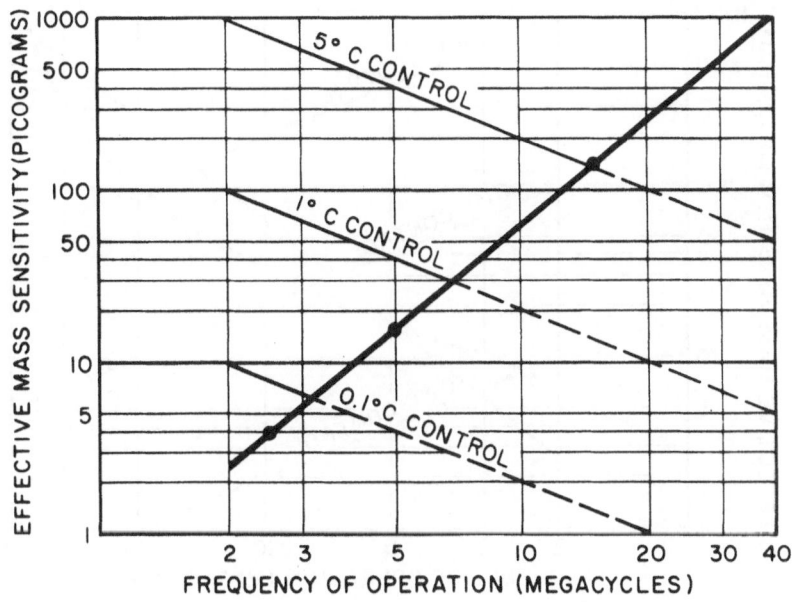

Fig. 4. Limiting effect of ±1 db change of amplitude of vibration.

quartz plate is of the order of 10^{-7} cal/sec; further, the effect is apparently instantaneous. Even at these small strain amplitudes, the material is not perfectly elastic. Apparently, dislocations in the quartz contribute an additional strain component to the stress field.[4] The slope of the heavy line was determined experimentally from three points where careful measurements have been made in connection with the fabrication of frequency standard crystals.

EFFECT OF Q

It is quite probable that improvements in circuitry will reduce the limitation imposed by amplitude effects,

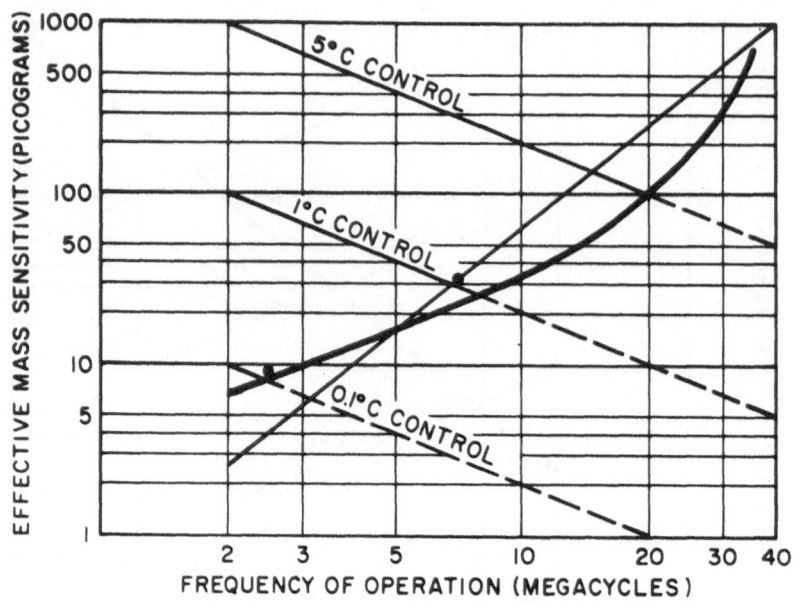

Fig. 5. Limiting effect of crystal Q and impedance level.

but nearly paralleling it is another limiting factor, the Q of the quartz plate. The Q of natural quartz in the neighborhood of room temperature varies inversely with frequency,[5] while the requirement for Q (to compensate for changes in cables and other sources of phase instability in the oscillator circuit) increases with frequency. Hence, in spite of the increased frequency sensitivity to added mass at the higher frequencies, the E.M.S. is affected adversely, as shown in Fig. 5. An additional parameter is involved in that the impedance of the quartz plate is excessively low at the higher frequencies. The effective Q at the circuit is less than that of the quartz plate, due both to the resistance of the cable and the difficulty of making a correct impedance match to the circuit. This factor accounts for the steeper rise of the curve above 10 Mc. By way of example, it can be seen from Fig. 5 that if ±1° C temperature control were feasible and a 7-Mc plate were used, then mass changes could be measured with an E.M.S. of about ±30 picograms. If ±0.1° C control were possible, the choice would be a 2.5-Mc plate with an E.M.S. near ±10 picograms.

Operation at still lower frequencies would appear even more attractive. However, it requires inordinately large plates to maintain the intrinsic Q of the quartz. With a lowered Q, the frequency of resonance is poorly delineated and the mass correspondingly uncertain. Even in the 2.5- to 20-Mc range, quartz plates must be contoured in order to localize the vibration to the center of the plate, away from the mounting points. At 5 Mc the smallest plate that realizes the Q of quartz, $3 \cdot 10^6$, is 15 mm in diameter. At 2.5 Mc it is 30 mm, and doubles each time the frequency is halved.

OVERTONE MODE OF OPERATION

One solution to this dilemma of size, particularly when the temperature control can be improved, is to operate the quartz plate in an overtone mode of vibration. Figure 6 shows the effect of using the fifth harmonically related overtone. For a given frequency, the plate is five times as thick, but the impedance becomes 125 times higher. By sacrificing the frequency sensitivity to mass by a factor of five, the precision of frequency measurement is increased an order of magnitude or more. The result is a reduction in E.M.S. by a factor of 2 when the effective Q factor is dominant. With this ap-

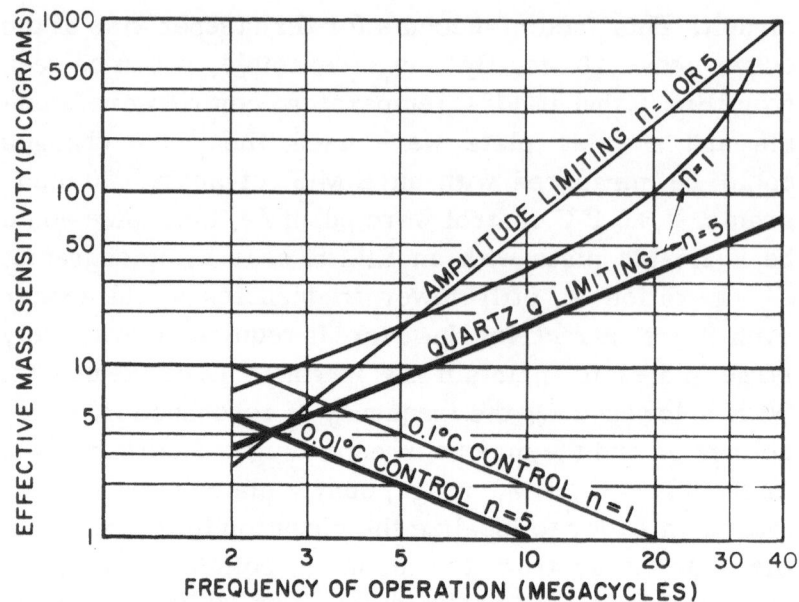

Fig. 6. Use of the overtone mode to improve E.M.S.

proach it appears likely that a mass change of about 1 picogram on an area of 10 mm^2 can be detected.

TYPICAL CONSTRUCTION

Figure 7 shows one of the crystal plate assemblies used for micro-weighing. The quartz surface is polished, both for best operation as an oscillator and for a more accurate estimation of the surface area. Gold is chosen as the electrode material for its physical and chemical stability. The support wires are fastened to the quartz plate by thermocompression bonding,[6] pass through a glass supporting platform, and are brazed to the platform rods. The assembly may be baked at 300° C. This crystal

Fig. 7. Crystal plate assembly, which may be baked at 300°C.

plate has a fundamental resonance at 5 Mc with a Q of over one million. Its capacity for additional mass is estimated at about 2000 μg, which can be measured to about 20 picograms.

THERMAL EFFECTS

Quartz plates, particularly the thicker ones, exhibit for a period of time a much larger frequency excursion than is predicted by their equilibrium temperature coefficient, whenever the ambient temperature is changed. Figure 8 shows the effect of a rapid change in ambient of 0.5° C. Only the residual change (B) of 10 parts in 10^9 is due to the temperature coefficient. The larger changes (A and C) of 50 parts in 10^9 are attributed

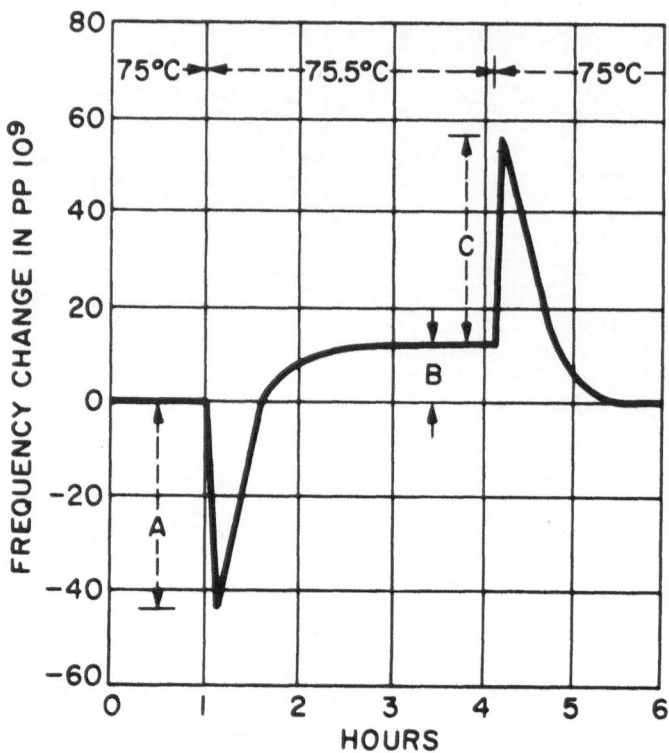

Fig. 8. Effect of thermal shock on a 3-mm-thick quartz plate.

to the effect of temperature gradients within the quartz plate.[7]

To obtain the plot in Fig. 8 the crystal plate was enclosed in an evacuated glass envelope and mounted in a cylindrical brass oven. The effect of increasing the temperature of the brass cylinder 0.5° C was to lower the frequency 50 parts in 10^9 in about 3 minutes. It took over an hour for the frequency to recover from thermal gradients and to arrive at a new stable equilibrium frequency about 10 parts in 10^9 higher as predicted by the equilibrium temperature coefficient. Reversing the direction of the temperature reverses the direction of frequency change.

This effect is quite reproducible, and when conditions are right for frequency measurements to parts in 10^{10}, extremely small thermal shocks at the surface of the quartz plate can be detected. The character of these spikes is such that they can usually be recognized and so do not interfere with a reading of the change in mass.[8]

OSCILLATOR CIRCUIT CONSIDERATIONS

The two limitations to micro-weighing with quartz plates related to circuitry are amplitude control and phase stability. Some progress has been made in improving amplitude control by use of temperature-controlled transistor amplifiers,[9] but basically the difficulty is that the resonant frequency of extremely thin high-frequency quartz plates is too sensitive to the amplitude output of the driving oscillator to make good mass change detectors.

At 5 Mc and below, this amplitude limitation becomes very small and is less than that caused by phase instabilities in the oscillator. Phase instability arises from variations in any component in the oscillating loop, especially capacitance changes in the connecting cables. In general, the phase stability is a direct function of the Q of the quartz plate and an inverse function of the frequency of operation.

TYPICAL OSCILLATOR CIRCUIT

Figure 9 shows the circuit, cover removed, used with the 5-Mc unit of Fig. 7. This is actually two electrically separate circuits which permit experiments

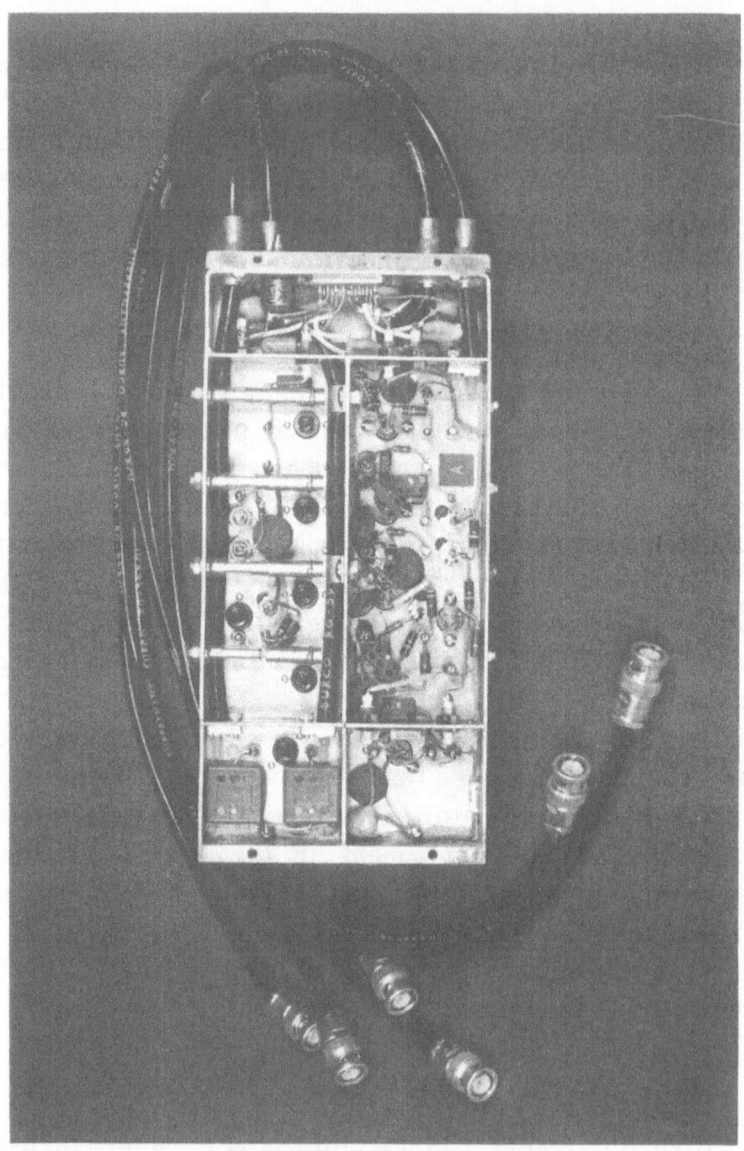

Fig. 9. Duplex circuit for use at 5 Mc.

where it is desirable to balance out certain ambient variations which are coherent in two crystals, one being used as a reference. The circuits are identical, and the top of one and the bottom of the other can be seen. The single high-gain transistor oscillator is visible at the left, as are the large terminating capacitors which minimize cable effects. The five-transistor amplifier and drive level control circuit is in the larger compartment. An additional transistor circuit, not shown, is used to regulate the temperature of the entire assembly to $\pm 0.1°C$. The oscillator and amplifier schematic circuit is shown in Fig. 10.

FREQUENCY MEASUREMENT

The E.M.S. is limited by the accuracy of frequency measurement in much the same way as by temperature control, since both are in terms of a percentage change in frequency. The requirement for accuracy in frequency comparison can be determined from Figs. 3 or 5. The line for 1° C control requires an ability to measure frequency to at least 1 part in 10^8; for 0.1° C in control, 1 part in 10^9; and so forth. Fortunately, equipment is available to make and record the necessary frequency measurements.

For measurement accuracy of $3 \cdot 10^{-10}$ g/cm^2, which can be made near 7 Mc, all that is required is a frequency measurement accuracy of 1 part in 10^8. This can be done with an electronic counter averaging for 10 sec, that is, counting cycles directly for 10 sec. The internal reference standard may be used with an occasional check against a primary frequency standard such as station WWV or NBA.

MICRO BALANCE OSCILLATOR

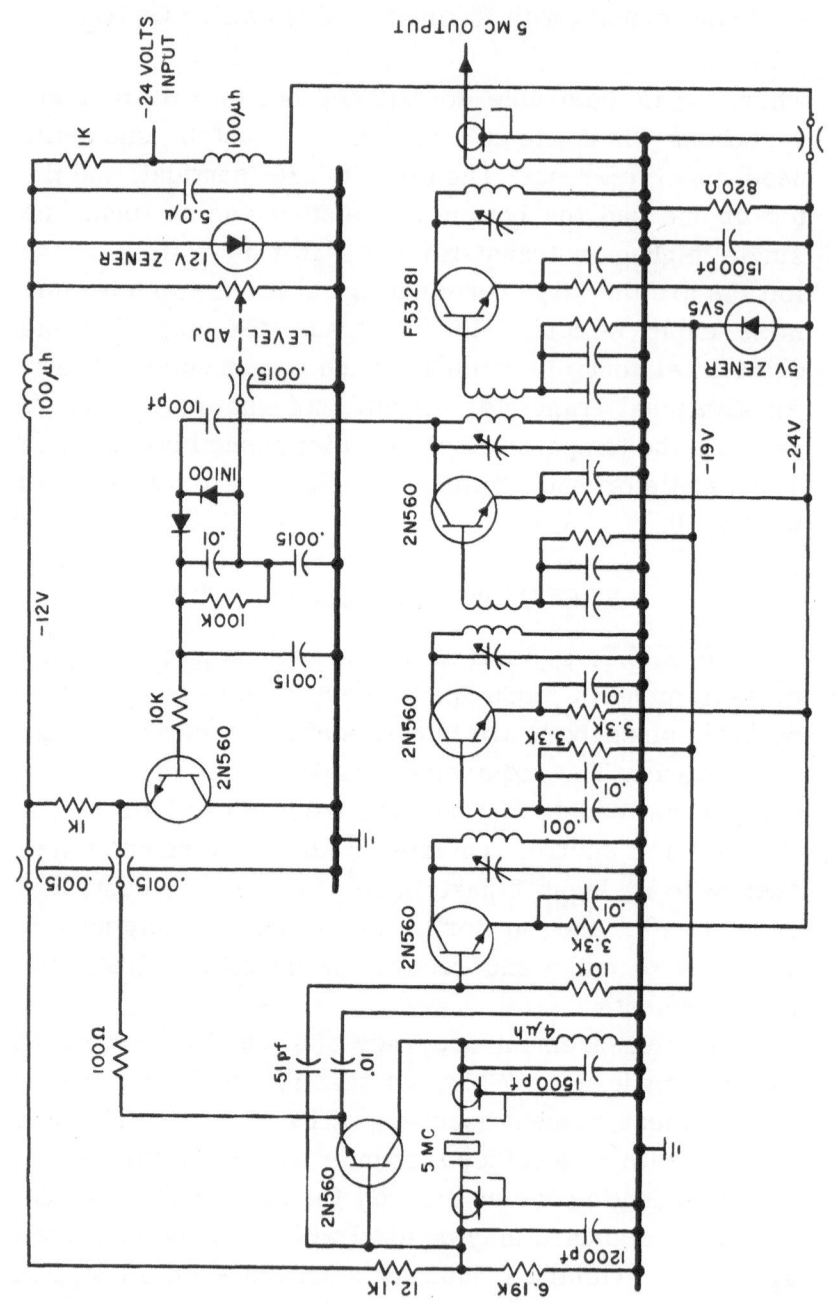

Fig. 10. Microbalance oscillator.

For measurement accuracy of 5 picograms (50 pico-grams/cm^2), which would necessarily involve thicker, less mass-sensitive quartz plates, frequency measure-ments (and temperature control) two orders of magnitude better are required. Such accuracy of 1 part in 10^{10} is accomplished as follows. The basic standard is station NBA in the Canal Zone, which broadcasts at 18 kc with an accuracy of a few parts in 10^{11}. This in turn is used to monitor a local oscillator at 2.5 Mc, which has a short time stability of 1 part in 10^{10} and a drift rate of about 2 parts in 10^{10} per month. The 2.5-Mc oscillator controls a commercial frequency synthesizer which will supply any frequency up to 30 Mc in 1-kc steps (or 1-cycle steps, with additional equipment). By use of period meas-uring techniques, the unknown frequency can be compared to the nearest reference frequency with the required accuracy.

The period measuring technique consists in meas-uring the beat note or difference frequency between two precise frequencies by measuring the number of micro-seconds between beats or some multiple of beats. The equipment uses an electronic counter set for period measurements. A frequency of 1 Mc provides the micro-second counting, and the low beat frequency is used to turn on and off the microsecond counter at zero crossing.

For example, given two 10-Mc signals about 100 cycles apart, a 1-cycle change (or 1 part in 10^7) will change the period about 1% or 10^4 μsec in 100 periods. Likewise, a change of one part in 10^{10} will change the time for 100 periods by 10 μsec. Therefore, the required frequency accuracy for an E.M.S. of 5 picograms can be achieved in 1 sec (time for 100 periods). Thus dynamic measurements can be made either of the mass effect or

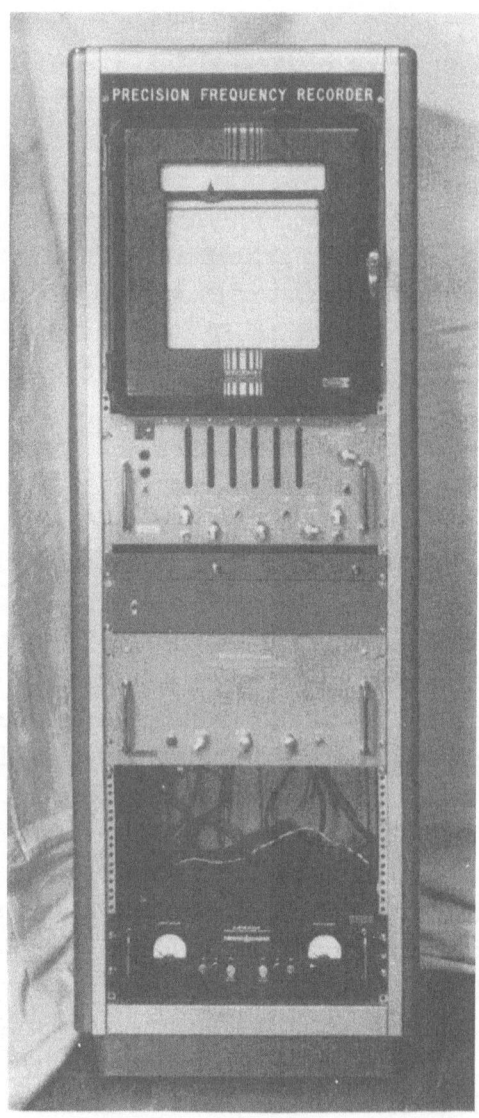

Fig. 11. Frequency comparison apparatus.

the thermal effect depending on which effect is dominant.

Some of the frequency comparison apparatus is shown in Fig. 11. Its advantages include the ability to make accurate comparisons between frequencies as much as 1000 cycles apart. It has a wide range of sensitivity and a long effective chart width. All the components except the modulator are commercially available.

CONCLUSIONS

The evidence suggests that micro-weighing by means of oscillating quartz plates can best be done at frequencies between 2.5 and 10 Mc. Although the sensitivity to mass change is greater at higher frequencies, the problems associated with the delineation of the quartz plate resonance more than offset the advantage.

Measurement with an accuracy of 50−100 picograms on a 10-mm^2 area can be made using a simple oscillator circuit, amplifier, and electronic counter. Measurements to a few picogram accuracy can be made if more sophisticated circuitry and temperature control are provided, and many experiments may now be considered which were not feasible before. Coupled to electron diffraction techniques, for example, the ability of a quartz resonator to detect masses commensurate with the minimum amount of material needed to define a structure pattern looks indeed attractive.

ACKNOWLEDGMENTS

The authors wish to acknowledge the assistance of J. C. King and J. P. Griffin in the design of the quartz plate mounting, and that of H. R. Beurrier, who designed and built the circuit, and J. R. Merchant, who processed and mounted the quartz plates.

REFERENCES

1. G. Sauerbrey, Z. f. Physik 155, 206-222 (1959).
2. A. W. Warner, "Frequency Aging of High-Frequency Plated Crystal Units," Proc. I.R.E. 43, 790 (1955); "Design and Performance of Ultra-Precise 2.5 Mc Quartz Crystal Units," Bell System Tech. J. 39, 1193 (1960).
3. R. A. Heising, Quartz Crystals for Electrical Circuits, D. Van Nostrand Co., New York, 1946.
4. A. Granato and K. Lücke, J. Appl. Phys. 27, 583 (1956); 27, 789 (1956); 28, 635 (1957).
5. H. E. Bommel, W. P. Mason, and A. W. Warner, "Dislocations, Relaxations, and Anelasticity of Crystal Quartz," Phys. Rev. 102, 64 (1956).
6. O. L. Anderson et al., "Technique for Connecting Electrical Leads to Semiconductors," J. Appl. Phys. 28, 923 (1957).
7. D. L. White, "An Ultra-Precise Standard of Frequency," 11th Interim Report. Dept. Army 30-039 sc 73078.
8. C. D. Stockbridge and A. W. Warner, this volume, p. 93.
9. W. L. Smith, I.R.E. Trans. on Instrum. I-9, No. 2, 1 (1960).

A VACUUM SYSTEM FOR MASS AND THERMAL MEASUREMENT WITH RESONATING CRYSTALLINE QUARTZ

C. D. Stockbridge and A. W. Warner
Bell Telephone Laboratories, Inc.
Whippany, New Jersey

ABSTRACT

A vacuum system for mass and thermal studies utilizing a resonating crystalline quartz plate is nearing completion in our laboratory. This system features some of the most recent aspects of ultrahigh-vacuum technology, among which are (1) a glass envelope with metal valve, an ion-getter pump, and a metal O-ring seal, (2) a cold-cathode magnetron total pressure gauge linear to 10^{-12} torr, and (3) a magnetic deflection mass spectrometer for partial pressure analysis of the crystal ambient. The system includes an inverted immersion oven for $\pm0.01°C$ control of the quartz plate and is bakeable to $450°C$.

Studies proposed for this apparatus include (1) determination of the limit of frequency stability of quartz resonators in analytical vacuum enclosures and, hence, the limit of minimum significant mass change possible, (2) analysis of gases evolved on fracture of quartz, (3) determination of the rate of transport of metals around vacuum systems by trace halogens, and (4) analysis of monolayer sorption phenomena by the thermal shock to quartz.

Some preliminary results with a simpler vacuum system are reported which make possible a rough calculation of the heat of adsorption of approximately $1/40$ monolayer of dry air on gold.

INTRODUCTION

In sequel to the previous paper[1] on the selection and design aspects of quartz crystals for mass change sensing, we now describe a vacuum system whose express purpose is to realize the extreme sensitivity of suitable quartz crystals to minute changes in quantity or position of mass on their surfaces and, further, to attempt utilization of the thermal phenomena which accompany these changes. Only with such apparatus will it be profitable to study the effects of time on the crystal ambient and vice versa to examine the effects of ambient changes on the resonator frequency.

Design Considerations

As a logical starting point we begin our description with the crystal plate and work out to the atmosphere, mentioning briefly the ambient measuring equipment and pumping facilities en route.

The Quartz Crystal Plate

Design began with the decision to use an edge-mounted AT-cut quartz plate vibrating in thickness shear of the type currently used in many primary frequency standards.[2] The quartz is polished planoconvex to confine or limit the piezoelectrically excited mechanical vibration to the central volume of the circular plate, thereby leaving the edge quiescent. The virtue of such crystal plates is their relative frequency insensitivity to the dimensions and physical properties of the mounting wires, pins, or ribbons.

To excite and maintain acoustic vibration of the quartz plate either a parallel or a perpendicular alternating electric field may be applied.

Parallel-Field Excitation

Excitation of the quartz crystal by a parallel field leaves the mass-sensitive surfaces bare, that is, not covered by a metallic or other electrically conducting thin film. However, little is known about the technique, which has not yet been exploited to any great extent; some details are given by Bechmann.[3]*

Perpendicular-Field Excitation

More usually, an electric field applied perpendicular to the direction of motion of the excited quartz lattice is used. A conducting electrode overlies each of the mass-sensitive quartz surfaces and becomes an integral part of the resonator. We do not yet know the difference, if any, between the radial sensitivity curves for added mass on the plane and the convex sides.[4]

As in primary frequency standard practice, we are presently using perpendicular-field excitation applied between vacuum-evaporated pure gold electrodes of between 500 and 1000 A thickness. These electrodes,

*V. Ianouchevsky, Observatoire de Paris, is also working with parallel field excitation of quartz.

though delicate and weakly held to the quartz surface by van der Waals forces,[5] have proved the most stable at low strain amplitude. Reactive metal electrodes which adhere strongly to the quartz, such as aluminum or iron,[5] can and have been used, but with these chemically bound metal films, difficulties with oxidation and incomplete residual stress relaxation must be anticipated.

Crystal Mounting

Since there are no requirements of mechanical ruggedness in terms of ability to withstand the vibration of transportation, in the present design we elected to support the quartz plate in a horizontal plane by a three-point gravity mounting. Thus, three short, tapered gold pegs forced through ultrasonically drilled holes near the periphery of the 30-mm-diameter quartz plate support the crystal on a glass platform within the vacuum enclosure. This is illustrated in Fig. 1. Two of the support pegs are in electrical contact with the electrodes; the third is for use with a bimetallic thin-film thermocouple evaporated onto the quartz plate.

The Crystal Oven

It was pointed out in the previous paper[1] that control of the quartz crystal temperature is one of the most important ambient parameters affecting the frequency of resonance. Thus from the iso-delta-temperature plot shown in Fig. 3 of Ref. 1, if we require that the error in frequency of resonance expressed as picograms for a 5-Mc AT-cut resonator oscillating at its frequency temperature turnover point should not exceed 40 picograms, then variation of the temperature of the crystal (varied infinitely slowly) must not exceed ±1° C. In fact, much better control is desirable, to ±0.01° C, in order to reduce the probability of a rapid temperature change of less than 1° C, which would thermally shock the crystal into a large frequency excursion.

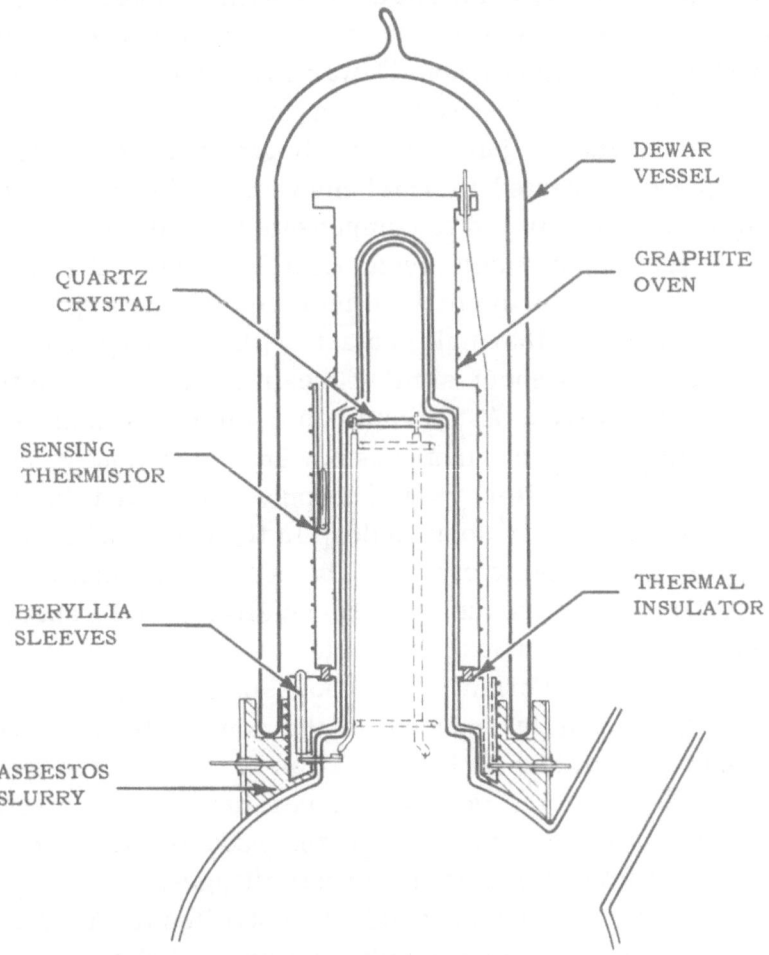

Fig. 1. Inverted immersion quartz crystal oven showing crystal resting on pins in a horizontal plane.

Such temperature control can be practically achieved with a well-designed fluid (water) bath (<±0.001°C) or with a solid oven totally enclosed within a second solid oven. Bakeout and operation at temperatures above 100°C are possible with the solid ovens, and this type was chosen for our system. The design is illustrated in Fig. 1, which shows the crystal immersed in an inverted solid oven under a high-vacuum Dewar filled with quartz wool. This oven is necessarily open at one end where the crystal envelope joins the analytical vacuum system; a separately heated thermal mass near the oven mouth has been incorporated to compensate for radiation losses from the upper main oven. Heat loss down the wires from the crystal plate is minimized by keeping these wires in good thermal contact with the oven over an appreciable distance; similar treatment is accorded the thermistor wires. As is common in quartz crystal oven practice, the oven winding is powered by a continuously variable (transistor) power supply controlled by the sensing thermistor[6] buried deep in the oven. Aluminum, copper, and graphite ovens are being tried pragmatically.

The salient virtues of the small-thermal-capacity oven with a continuously variable power input are (1) the elimination of thermal overshoot, since the power input is proportionately reduced as the control temperature is approached, and (2) the reduction of the time the temperature is off the desired temperature in response to an ambient temperature change; this reduction is achieved without introducing abrupt thermal changes.

Working against an ambient controlled to ±1°C, this oven easily achieves an ambient ratio of 100 or regulation to ±0.01°C. Selection of the oven operating temperature depends on the precise orientation of the quartz crystal lattice to the effective plane of the flat side of the particular crystal plate under test. In practice the procedure is to attempt preparation of a quartz plate with a calculated temperature vs. frequency turnover point about 10° C above room temperature ambient, and then, having prepared a resonator, to find empirically the temperature of the actual turnover point.

Total Pressure Measurement

The requirements of our system in terms of pressure measurement are that the pressure-sensitive device should:

1. exert a minimum disturbance on the ambient and, hence, on the crystal,
2. have a high sensitivity at 10^{-10} torr,
3. have no hot filament, both because of its thermal radiation and the chemical reactions[7] involved such as $H_2O + C$ (tungsten impurity) $\rightarrow CO + H_2$.

The Redhead cold-cathode magnetron gauge[8] meets these requirements quite well. This gauge has a sensitivity of 4.5 amp/torr and is linear down to at least 10^{-12} torr. It has no x-ray limit and operates with an anode potential of 6 kv and a magnetic field of 2000 gauss. The ion current, which is linearly proportional to the total pressure, is measured between the cathode and auxiliary cathode, which operate at room temperature.

Partial Pressure Measurement

We are also including facilities for partial pressure analysis of the ambient with a bakeable mass spectrometer. To begin with, the 180° deflection magnetic Diatron 20 C unit is being used; later a more sensitive prolate cycloidal crossed-field type[9] with a partial pressure sensitivity of $5 \cdot 10^{-12}$ torr will be fitted.

Location of the Ambient Analytical Tools

To increase the correlation between events occurring on the quartz plate electrode surface and in the analytical appendages these tools should:

1. have entrance apertures pointing in an opposite direction to the pump mouth, and
2. point directly at the surface of the quartz plate, since the mean free path of species leaving the plate will be many times the dimensions of the vacuum system in ambients whose "monolayer time" is extended.

Vacuum Systems

The four requirements: (1) gravity support of the quartz plate, (2) immersion of the quartz plate in a bakeable oven, (3) entrance apertures of the analytical tools which face toward the quartz plate and away from the pump mouth, and (4) operation under a reasonably sized Dewar with a maximum of thermal symmetry, led to the design shown in Fig. 2. This design, however, is too elaborate to construct in glass, particularly as a first venture. We have, therefore, in fact, reluctantly sacrificed requirement 3 and built the apparatus shown in Fig. 3. In this system a Bayard-Alpert gauge for pressure measurement during bakeout, a Redhead gauge, and mass spectrometer head are symmetrically disposed about and below the crystal oven base.

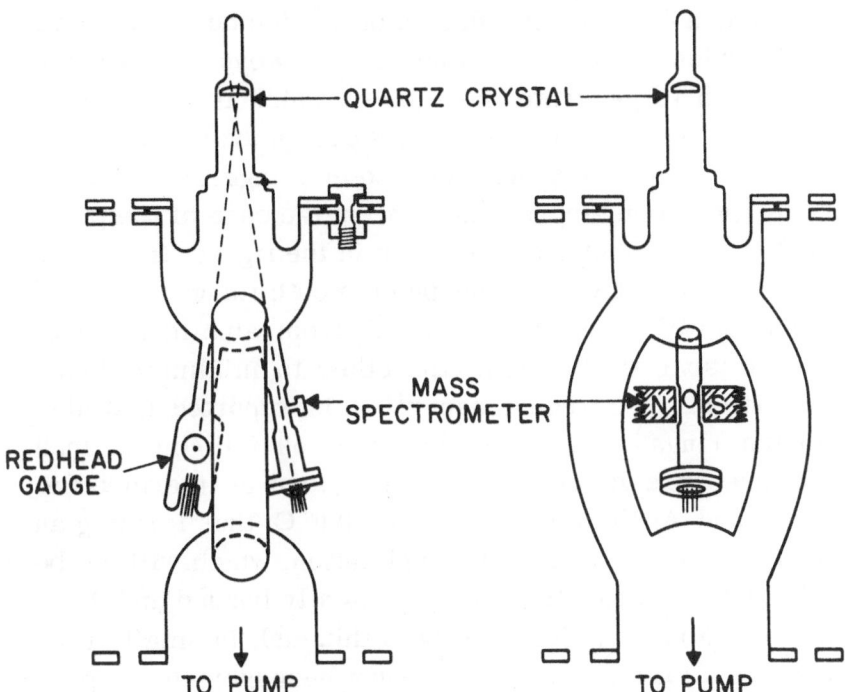

Fig. 2. Vacuum system with most desirable features.

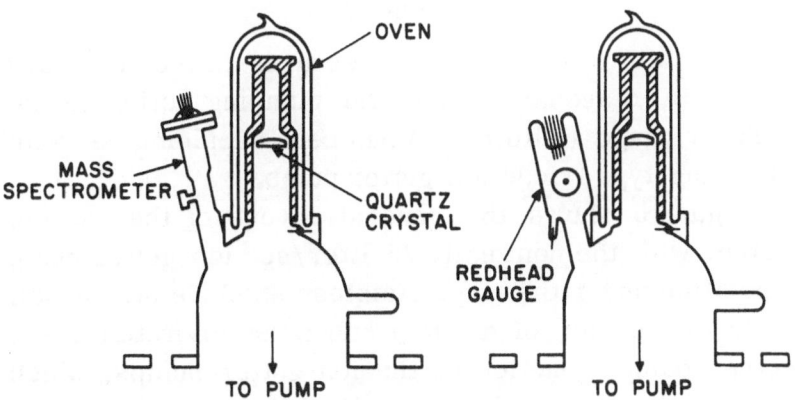

Fig. 3. Vacuum system with threefold symmetry (the Bayard-Alpert gauge is not shown).

Figure 4 shows the design of the demountable metal seal, which we believe is novel. Flat Kovar sheet stock was spun to generate the tube to which the 7052 glass envelope has been glassed. This design avoids the usual weld to a massive stainless steel flange and its concomitant outgassing properties. In the case of the pump seal, which is not the one shown in the figure, the Kovar lip butts directly onto the pump mouth flange using two aluminum O-rings "Koldwelded" from aluminum wire, one to seal the vacuum, the other to minimize flange distortion. In a recent paper[10] it is reported that aluminum rings, plastically flowed by 5000 lb/linear inch bolt pressure at room temperature, form an intermetallic compound Al_3Fe when heated to 400° C, thus forming an ideal vacuum seal for this application, which will not be demounted frequently. The chemically bonded and thermally stable seal fortunately is thin and, in small cross section, cleaves readily. It is not necessary to prepare the flange surfaces each time the seal is remade.

Pumping

In order to achieve low pressures in the 10^{-10} torr decade in reasonable time the pumping action of the Redhead gauge (1 liter/sec) has been supplemented with a Penning-type triode ion-getter pump.[11]

Figure 5 shows the general layout of the vacuum system with the nominally 25-liter/sec ion-getter pump in position under the large stainless steel Dewar vessel.

In this choice of a pump one must tolerate the selective pumping action of the ion-getter pumps, which is somewhat reduced in the Brubaker triode design,

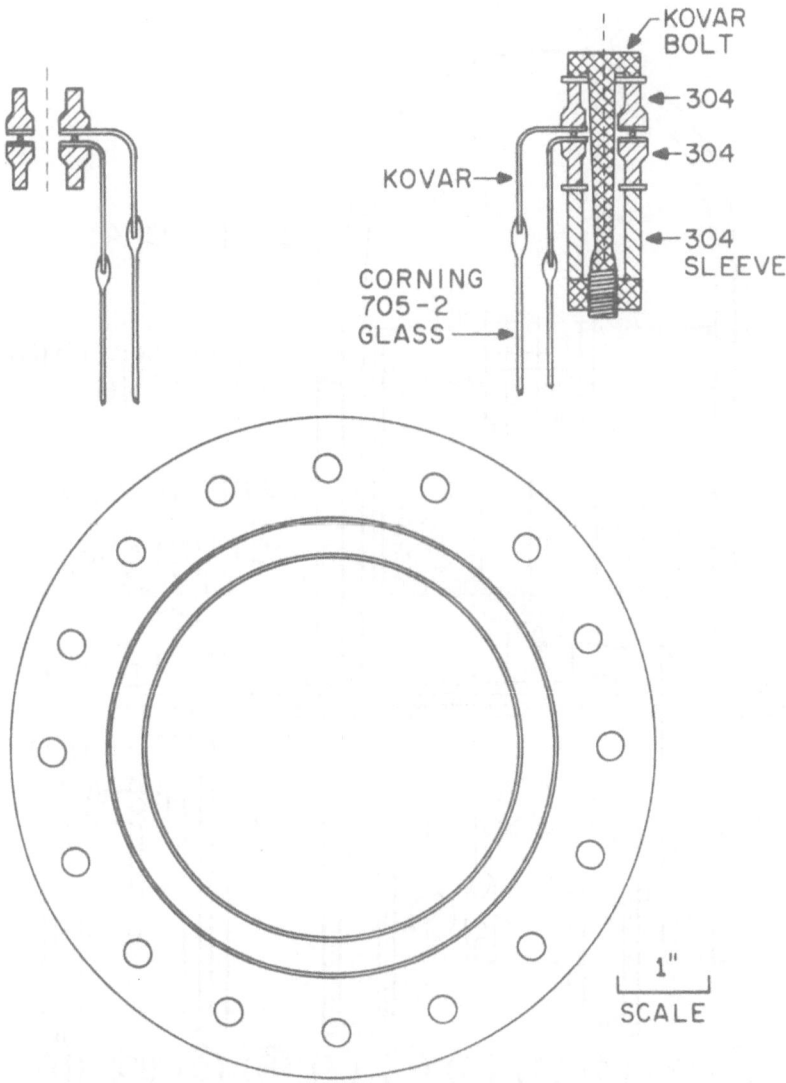

Fig. 4. Kovar seal; the larger half of this demountable joint is used to seal
the vacuum enclosure to the pump flange.

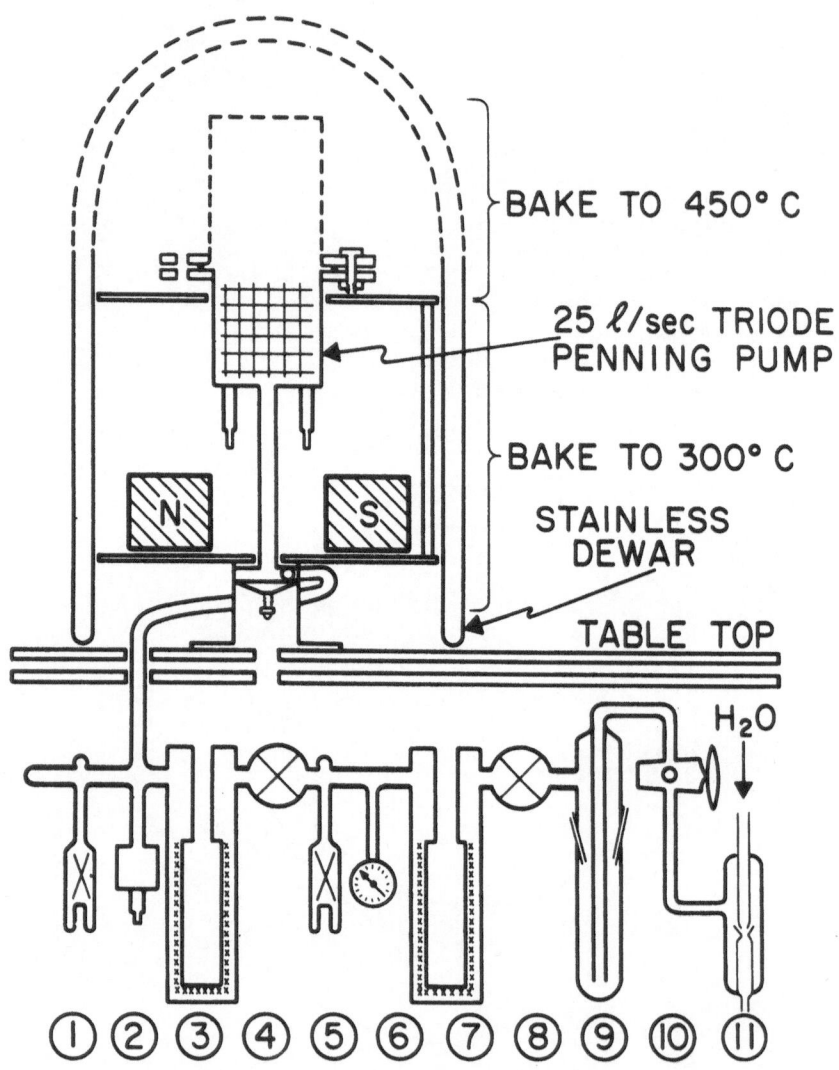

Fig. 5. Complete vacuum system; for identification of numbers, see text.

since these pumps have outstanding virtues for quartz crystal work, including the fact that they

    1. are free from oils and their decomposition products; these hydrophobic compounds are one of the persistent evils of present-day quartz crystal units since the frequency of oscillation is highly sensitive to the adsorption and desorption of partial monolayers,

    2. do not require isolation traps,

    3. can be baked out at moderately high temperature,

    4. can easily be turned on and off.

To facilitate construction, the whole assembly is mounted by the exhaust tubulation to an all-metal bakeable silver-seat valve.[12]

Below the bakeout table the various components are: 1) thermocouple gauge R.C.A. 1946, 2) diode ion-getter pump for bakeout pumping and pressure measurement, 3) internally bakeable (300° C) persorption pump containing Linde 13X molecular sieve (the 4A and 5A molecular sieves are not suitable, since there is a decrease[13] in the pumping speed for nitrogen below −120°C), 4) all-metal valve, 5) thermocouple gauge, 6) Bourdon-type gauge, 7) internally bakeable persorption pump, 8) all-metal valve, 9) water vapor trap, 10) P.T.F.E. plug stopcock, and 11) water aspirator.

With this system we shall (1) determine the frequency/time stability limit or lowest possible drift rate of quartz crystals in an analytical vacuum system, (2) investigate the electrolysis of quartz at moderate temperatures and determine what gases, if any, are released at the anode, and (3) look into the transport of halides around a vacuum system. Here we have in mind finding out if there is a mechanism by which trace halogens can effect the distribution of metals more or less uniformly over the interior surface of an isothermal vacuum enclosure.

RESULTS AND DISCUSSION

To conclude and illustrate the potentialities of the quartz crystal as a mass-sensing device we shall describe an interesting effect obtained with a similar but simpler system. The apparatus consisted of a fifth-overtone 2.5 Mc AT-cut quartz plate resonated in an isothermal environment at about 43° C to ± 0.01° C stability and connected via a narrow tubulation to a Bayard-Alpert gauge and a 1-liter/sec diode ion-getter pump.

Figure 6 shows the effect of admitting clean dry air to this system, which had previously been 'baked' at 105° C for 3 days. At 43° C the crystal operated with a frequency stability of 1 to 2 parts in 10^9 per day.

From a pressure of 10^{-8} torr to 10^{-3} torr no singular effect on frequency is noted, but from 10^{-2} to 10^{-1} torr there is a pronounced and partially recovering shift, to a frequency about 1 part in 10^9 lower. This occurred at approximately 50 μ. On exhausting the system, as the pressure fell through this range, an inverse shift of frequency occurred, this time to a frequency about $1\frac{1}{2}$ parts in 10^9 higher, with partial recovery to the original high-vacuum aging rate of 1 part in 10^9 per day. Very modest pumping speeds were required for this effect, nominally 1 liter/sec at the pump, much less at the resonator tubulation. This pattern of a reversible frequency shift was repeated several times and is not spurious.

Discussion

That the frequency shift on admitting air was simply due to the cooling effect of room-temperature air, possibly augmented by expansion at the tubulation orifice, can be dismissed as an explanation without analysis, since the effect was reversible, acted in the opposite sense to that required by the temperature coefficient of the crystal, and did not occur in the higher pressure range.

In Fig. 6 the shift of frequency, Δm, of about 1 part in 10^9 indicates that the resonator became covered with approximately the fortieth part of a monolayer of some gas, quite probably oxygen.[14] We feel there is a connection between this reversible sorption of gas on gold and the reversible change of 20 to 30 mv in the contact potential difference found by Nadjokov et al.[15] on taking gold electrometer electrodes in and out of a vacuum ambient.

The whole phenomenon looks like one of thermal shock, which was discussed in the previous paper[1] (see Fig. 8). We shall now analyze Fig. 6 in terms of a mathematical model.

A heat shock Q which disturbs the resonant frequency f of a quartz crystal must in the last analysis produce this change Δf by applying stresses to the quartz. These stresses effect either a change in dimension t, modulus of elasticity c, or density ρ. This is a direct consequence of the fundamental frequency equation,

$$f = \frac{n}{2t} \sqrt{\frac{c}{\rho}} \qquad (1)$$

describing a plate vibrating in a thickness mode (the AT-cut uses a thickness shear mode), where n is the

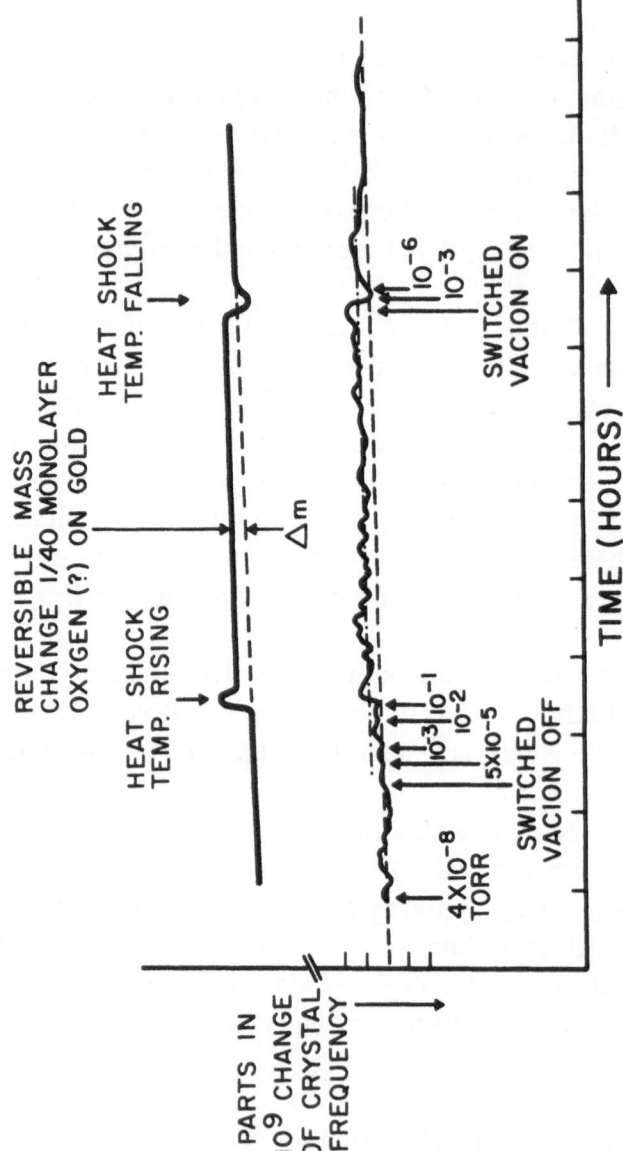

Fig. 6. Effect of condensing a partial monolayer of gas on a gold-covered quartz resonator; idealized frequency-time plot shown above a copy of the actual recorder trace.

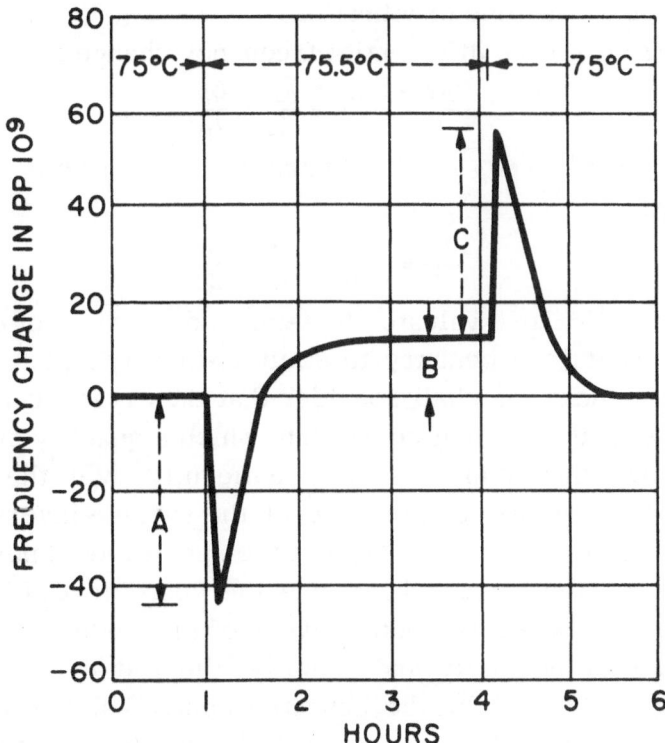

Fig. 7. Effect of an abrupt change of ambient temperature by 0.5°C on the resonant frequency of an AT-cut quartz plate. At time 1 hour the ambient temperature was changed from 75.0 to 75.5°C and held until time 4 hours 10 minutes, when it was returned to 75.0°C.

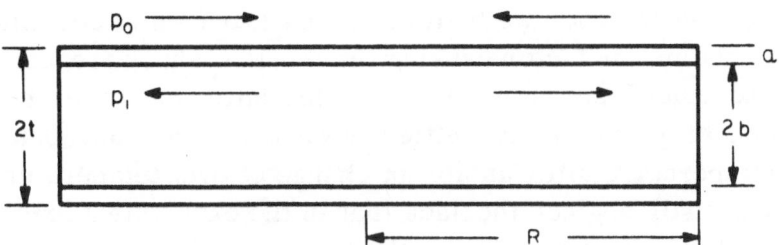

Fig. 8. Model for analysis of thermal shock to quartz by a condensing partial monolayer (see text).

harmonic overtone of vibration.

From (1) the differential frequency change is

$$\frac{\Delta f}{f} = -\frac{\Delta t}{t} + \frac{\Delta c}{2c} - \frac{\Delta \rho}{2\rho} \qquad (2)$$

and since $V = m/\rho = \pi R^2 t$, where R is the radius of the plate,

$$\frac{\Delta f}{f} = -\frac{\Delta t}{2t} + \frac{\Delta R}{R} + \frac{\Delta c}{2c} \qquad (3)$$

In order to evaluate the terms on the right of equation (3) it is necessary to elicit a simplifying but realistic model. We shall consider that the crystal consists of a circular slab of quartz into which a quantity of heat Q flows instantaneously and uniformly over the area πR^2, raising the temperature of the surface layers to a depth a by an amount ΔT. Let the specific heat per unit mass of the quartz be K and the coefficient of linear expansion be a. The outer layers of the model (see Fig. 8) expand normally and radially, the radial expansion being restrained by the unheated central slab of thickness $2b$; this places the central slab in tension. It is the resultant changes in thickness, density, and elastic modulus which cause the observed frequency change. At the present time we do not have the data to evaluate changes in the elastic modulus with strain, whose effect in the present instance we believe is small, but which certainly is the major factor causing the thermal shock frequency excursions shown in Fig. 7. In this latter case the whole quartz plate finally settles down to a new equilibrium temperature after an abrupt change of oven temperature. We shall neglect the fact that in the experiment expansion of the quartz beneath the electrodes was further restricted by an unheated annular ring of quartz.

For this model* we may write

$$\Delta R = p_i\left(\frac{1-\nu}{E}\right)R \tag{4}$$

$$aR\Delta T = \left[p_o\left(\frac{1-\nu}{E}\right)+ p_i\left(\frac{1-\nu}{E}\right)\right]R \tag{5}$$

$$\frac{p_o}{p_i} = \frac{b}{a} \tag{6}$$

where ν is Poisson's ratio, E Young's modulus of elasticity, and p_o and p_i are the outer and inner stresses shown in Fig. 8.

Solving (5) and (6) for

$$p_i = \frac{aa\Delta T}{[(1-\nu)/E]t}$$

and substituting in (4), we have:

$$\Delta R = \frac{aa\Delta TR}{t} \tag{7}$$

Now

$$\Delta t = aa\Delta T \tag{8}$$

so that substituting (7) and (8) in (3) gives

$$\frac{\Delta f}{f} = \frac{a}{2t}(a\Delta T) + \frac{\Delta c}{2c} \tag{9}$$

Δc may also be expressed in terms of $(a\Delta T)$, so that the solution to (3) may be written in any instance as

$$\frac{\Delta f}{f} = \beta(a\Delta T) \tag{10}$$

or, for a given $\Delta f/f$,

$$a\Delta T = \frac{1}{\beta} \cdot \frac{\Delta f}{f}$$

*We are indebted to H. F. Tiersten for the analysis of this model.

Now Q, the quantity of heat in calories causing the temperature rise ΔT in the outer layers of thickness a, is given by

$$Q = \frac{K}{\rho} 2\pi R^2 (a\Delta T) \tag{11}$$

Hence, substituting (9) in (11), we have

$$Q = \frac{K}{\rho} 2\pi R^2 \frac{\Delta f}{f} \cdot \frac{1}{\beta}$$

When the variation Δc with strain is neglected, $\beta = a/2t$ by (9). When Δc is negative, as will be the case for $Q > 0$ and the central slab in tension, then $\beta < a/2t$. Therefore, for a measured $\Delta f/f$, $(a\Delta T)$ is greater when $\Delta c/2c$ is included, and Q calculated is greater when Δc is included. Thus our answer given by

$$Q = \frac{K}{\rho} \frac{4\pi R^2 t}{a} \cdot \frac{\Delta f}{f} \tag{12}$$

is too small.

It is to be understood that Q given by (12) is the small quantity of heat which quasi-isothermally shocks the crystal; i.e., the final equilibrium temperature, when the stresses have fallen to zero, is unchanged.

Substituting experimental values $\Delta f/f = 1.5 \cdot 10^{-9}$ (from Fig. 6), $K = 0.2$ cal/g-deg, $\rho = 2.65$ g/cm^3, $R = 0.8$ cm, $t = 0.167$ cm, and $a = 10^{-5}$ deg^{-1} in (12), we have

$$Q = 1.5 \cdot 10^{-5} \text{ cal}$$

In order to estimate the net heat of adsorption it remains to calculate the mole fraction of gas (presumed molecular oxygen) causing this heat evolution Q. We can evaluate the mass change of the crystal on adsorption from the frequency shift shown as Δm in Fig. 6, after the thermal shock stresses had abated. With the particular crystal used the observed frequency change Δm of 1 part in 10^9 was equivalent to a mass change of $4.6 \cdot 10^{-10}$ g/cm^2. If this matter is uniformly adsorbed over the area of the electrodes, as was assumed in calculating Q, then it is not necessary to know the radial sensitivity curve of the quartz plate to added mass, and multiplication by the area of the electrodes gives the total mass adsorbed as $2.2 \cdot 10^{-9}$ g. The calculated heat of adsorption was, therefore, 109 kcal/mole, a value which, as we know from the literature,[14] is excessive. This is not surprising, considering the poor resolution of the frequency changes due to Q and Δm shown in Fig. 6.

The phenomenon is, however, sufficiently reproducible, sensitive, and reversible to merit further study. Empirical calibration with gases whose monolayer heats of adsorption are well known would seem appropriate before unknown systems are investigated.

REFERENCES

1. A. W. Warner and C. D. Stockbridge, this volume, p. 71.
2. A. W. Warner, Bell System Tech. J. 39, 1193 (1960).
3. R. Bechmann, Proc. I.R.E. 48, 367, 1278 (1960).
4. G. Sauerbrey, Z. f. Physik 155, 206 (1959).
5. P. Benjamin and C. Weaver, Proc. Roy. Soc. 254A, 177 (1960).
6. W. L. Smith, I.R.E. Trans. on Instr. I-9, 1 (1960).
7. J. R. Young, Proc. 19th Conf. on Physical Electronics, M.I.T. 1959.

8. P. A. Redhead, Proc. 19th Conf. on Physical Electronics, M.I.T. 1959.

9. E. V. Kornelsen, Proc. 19th Conf. on Physical Electronics, M.I.T. 1959.

10. J. Holden, L. Holland, and L. Laurenson, J. Sci. Instr. 36, 281 (1959); see also L. Holland, Seventh National Symposium on Vacuum Technology Transactions 1960, p. 168ff.

11. W. M. Brubaker, Vacuum Symposium Transactions 1959, p. 302.

12. S. C. Brown and J. E. Coyle, Rev. Sci. Instr. 23, 570 (1952).

13. D. W. Breck and J. V. Smith, Scientific American (January, 1959).

14. D. Brennan, D. O. Hayward, and B. M. W. Trapnell, Proc. Roy. Soc. 256A, 81 (1960).

15. G. Nadjokov, V. Vassilev, and S. Balabanov, C. R. Acad. Bulg. Sci. 11, 461 (1958).

ADSORPTION ON QUARTZ SINGLE CRYSTALS

W. H. Wade and L. J. Slutsky*

University of Texas, Austin

ABSTRACT

Preliminary measurements of the effect of the adsorption of gases on the resonant frequencies of quartz crystals have been made. Approximate adsorption isotherms for hexane and water vapor on Y-cut quartz crystals are reported.

INTRODUCTION

Traditionally, studies of physical adsorption are undertaken with the object of obtaining quantitative information about the energy of interaction between molecules of the adsorbed species and the solid phase. In addition, by a careful analysis of adsorption isotherms, or by measurement of the differential heat of adsorption as a function of surface coverage, the character of the lateral interactions between molecules of the adsorbed species may in favorable cases be elucidated. However, for comparison with a detailed theoretical model, data on adsorption by an unobstructed crystalline substrate

*Present address: Department of Chemistry, University of Washington, Seattle, Washington.

of known structure are necessary, whereas the limited sensitivity of conventional methods of measurement requires the use of a powdered sample. Thus, owing to surface distortion in sample preparation, irreversibly adsorbed contaminants, the presence of a variety of exposed crystal planes, and the effects of sample porosity, the solid—vapor interface is often so poorly characterized that it is not possible to interpret the data in terms of a microscopic model with any degree of confidence.

Recently, Sauerbrey[1] has pointed out that, in the case of a quartz crystal plate vibrating in a thickness shear mode, the elastic strain at the principal faces is zero at resonance. Thus, the deposition of a film of foreign matter at the surface should affect the resonant frequency of the crystal only as added mass and not through alteration of the surface elastic properties. Since crystal oscillators of great stability may be constructed and since the frequencies of such oscillators may be measured with considerable accuracy, this effect can provide a sensitive measure of the mass adsorbed on a quartz face, or on films deposited on quartz crystals. In this paper an apparatus for the study of the effect of the adsorption of water and hexane on the resonant frequencies of quartz plates vibrating in extensional and shear modes is described and some preliminary results are reported.

EXPERIMENTAL TECHNIQUE

The vacuum system used in this work is illustrated schematically in Fig. 1. The principal components are a thermostatted sample chamber which houses the quartz crystals, a wide-bore (15 mm) mercury manometer, and a bulb containing the liquid adsorbate. The system is largely glass. The valves (Hoke Type 413) and the stainless steel flanges sealing the sample chamber are soft-soldered to Kovar-to-glass seals. A three-stage oil diffusion pump, which is isolated from the working system by a liquid-nitrogen trap, produces pressures less

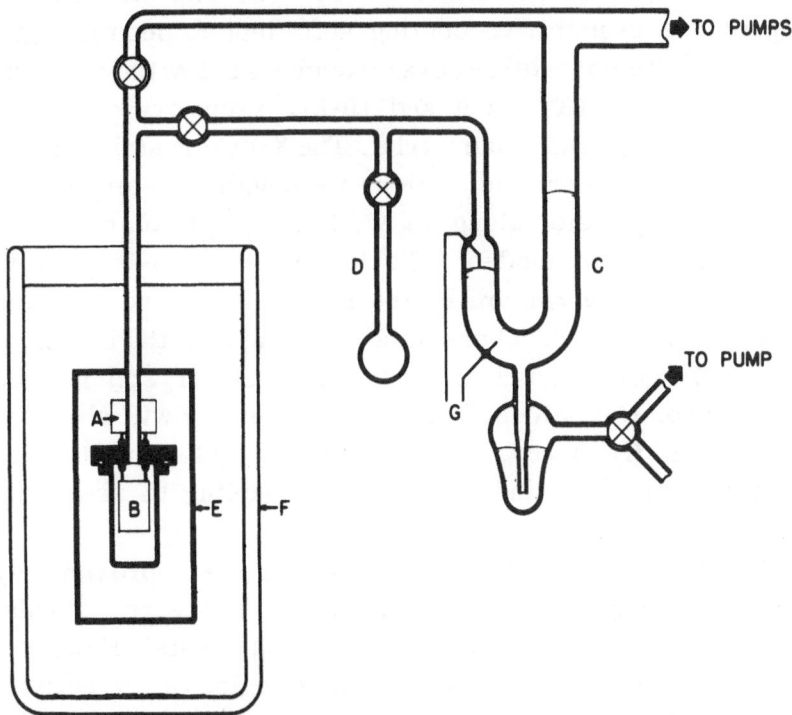

Fig. 1. Vacuum system and experimental chamber. A and B are the mounting position of the transistor oscillator and the crystal holder. C is the manometer, D is the liquid sample, E is the brass housing forming a dead-air space around the cell and oscillator, and F is the water bath. G is the probe for maintaining a constant reference height for one leg of the manometer.

than 10^{-7} mm Hg without baking of the metal parts.

The sample chamber and the oscillator are enclosed in a brass jacket submerged in an air-stirred water bath. The bath temperature is controlled to within $\pm 10^{-3\circ}$ C by a mercury regulator which activates a 10-w heater. However, since the vacuum chamber and the dead-air space within the submarine jacket constitute an appreciable thermal resistance, the excursions in the crystal temperature do not exceed $4 \cdot 10^{-4\circ}$ C. The air stirring provides evaporative cooling sufficient to permit regulation at temperatures several degrees below the ambient.

100-kc, $-18°$ X-cut and 10-Mc Y-cut crystals were used in this preliminary trial. The X-cut crystal vibrates in an extensional mode along its length with slight coupling to extension along its width and negligible coupling[2] to face shear modes. The length of a 100-kc plate is approximately 2.5 cm. Plates 1 cm in width and 0.076 cm thick were used. There is a nodal point in the center of a rectangular X-cut plate at which the crystal may be clamped. Deviations from nodal clamping will alter the resonant frequency, increase the equivalent resistance of the crystal, and enhance the coupling to shear and flexure modes.

The Y-cut crystals are circular discs approximately 1 cm in diameter and 0.020 cm thick. There are no nodal points on the principal faces of such a crystal. However, the crystal may be supported in a region of low activity by making the diameter of the exciting electrodes less than that of the crystal and clamping at the rim.

Crystal holders suitable for extensional and shear crystals are shown in Figs. 2a and 2b, the attendant oscillator circuits in Figs. 3a and 3b. In both cases the

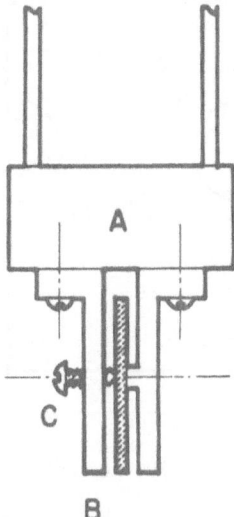

Fig. 2a. Holder for extensional crystals. The electrodes (B) are bolted to a flat glass plate (A). The crystal is clamped between a raised pedestal on the right-hand electrode and a pointed 0-80 screw (C).

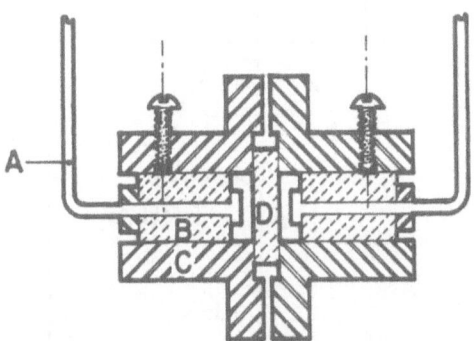

Fig. 2b. Holder for shear crystals. The crystal (D) is clamped between two stainless steel flanges (C). The electrodes (A) are insulated from the flanges by spacers (B) of 8-mm OD pyrex capillary tubing which are retained by stainless steel setscrews.

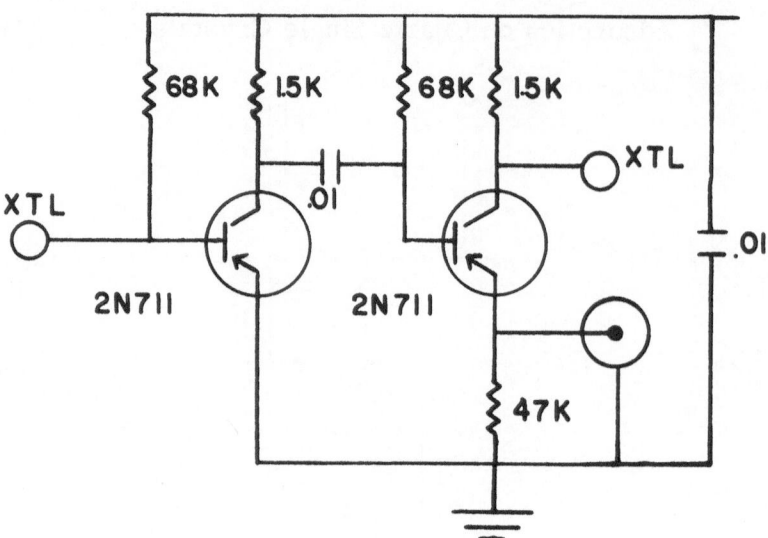

Fig. 3a. 100-kc oscillator.

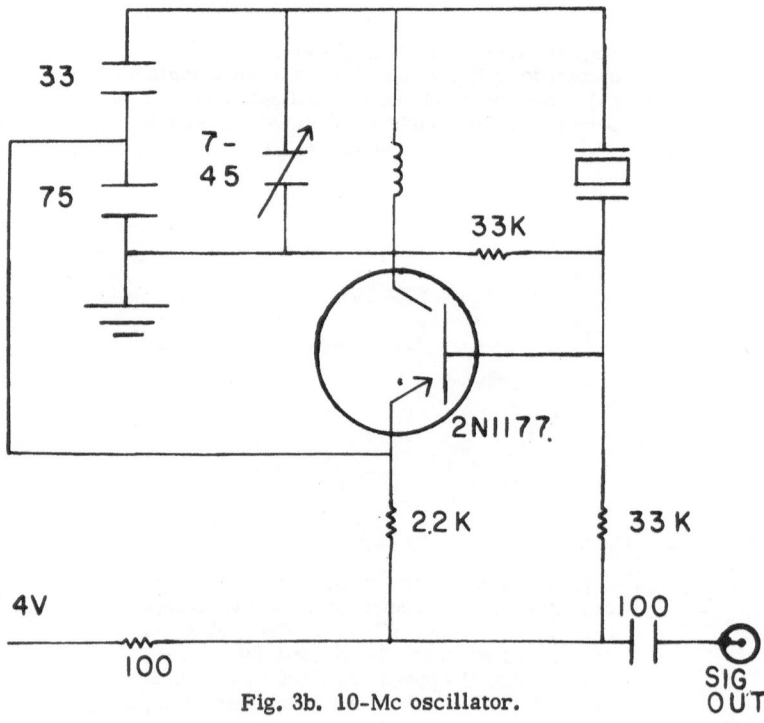

Fig. 3b. 10-Mc oscillator.

holders are fixed by setscrews to $\frac{1}{8}$-in. stainless steel rods which are taken through the top of the sample chamber by Kovar fittings. The oscillator is soldered to the upper ends of these rods. The mounts may be cleaned (if necessary after assembly) in chromic acid—sulfuric acid or hot nitric acid. Prolonged rinsing in frequent changes of warm distilled water is desirable to assure the removal of the last traces of cleaning solution from the screw threads and other crevices in the mount.

The setting of the gap between the electrodes and crystal is not critical. However, for the X-cut crystals it is probably desirable to use gaps of the order of $\frac{1}{8}$ wavelength to minimize possible effects due to acoustic resonance. In the case of the 10-Mc Y-cut crystals, a gap of approximately 0.003 in. was used.

Since the oscillators and batteries are rigidly mounted in a closely regulated environment, simple circuits without elaborate provision for amplitude and bias stabilization have been employed. The frequency drift rates were one part in 10^9 per minute or less. The frequency of the Y-cut crystal was measured by counting the oscillator output for a 10-sec interval with a Hewlett-Packard 524-C counter. The 100-kc signal from the X-cut crystal was multiplied up to 10-Mc before counting. The circuit shown in Fig. 3a may be used without modification for air-gap-mounted crystals with resonant frequencies up to 1 Mc.

EXPERIMENTAL PROCEDURE

The crystal assembly may be outgassed by an infrared lamp temporarily mounted outside the sample chamber. With an aluminum foil reflector taped to the opposite wall of the chamber it is possible to attain crystal temperatures in excess of 425° C. The lamp is controlled by a Powerstat and the voltage increased manually, the temperature being monitored by the change in resonant frequency. Since there is a possibility that an excessive thermal gradient may induce optical twinning,[3] the heating rates were in general kept below 3 deg/min. However, several X-cut plates heated at 8 deg/min showed no evidence of twinning.

Both the X- and Y-cut crystal have large temperature coefficients of frequency. Thus, if the temperature of the adsorbate gas is not very close to the temperature of the crystal, a transient shift in frequency is to be expected. This is particularly serious for the Y-cut crystal ($d\nu/dT$ = +90 ppm/deg), but is also quite noticeable for the −18° X-cut crystal ($d\nu/dT$ = −25 ppm/deg). Even if the adsorbate and crystal temperatures are closely matched, the thermal transient due to the heat of adsorption will generally be significant for the Y-cut crystal. In this work approximately two hours were required for the restoration of thermal equilibrium after the introduction of gas into the sample chamber. This long thermal relaxation time precludes the use of Y-cut crystals to study the kinetics of adsorption, but might make it possible, if the adsorbate gas were carefully equilibrated with the bath, to study heats of adsorption. The use of BT- or AT-cut crystals in place of the Y-cut would largely eliminate the difficulties associated with thermal transients.

The pressure of the adsorbate gas was measured with a traveling microscope readable to 0.01 mm. The water and hexane samples were degassed by repeated freezing, thawing, and evacuation. In addition, the hexane (Phillips Research Grade 99.9%) was dried over exhaustively outgassed molecular sieve.

RESULTS

The change in resonant frequency of a 10-Mc Y-cut crystal is plotted versus relative pressure of hexane vapor in Fig. 4. The 10-Mc Y-cut crystal has a mass of 0.0265 g per centimeter of geometrical surface area. Thus, a 1 ppm shift in frequency should correspond to $2.65 \cdot 10^{-8}$ g adsorbed per square centimeter of geometrical area, or 0.0706 cc adsorbed per square meter. A B.E.T. plot of the data in Fig. 4 implies a frequency shift at monolayer coverage of 2.5 ppm or an "area" of $20 \, A^2$ for the hexane molecule. This does not appear to be reasonable in the light of previous investigations by conventional volumetric adsorption techniques. If the value of $50 \, A^2$ found by Hackerman, Wade, and Every[4] for 0.9 m^2/g quartz is assumed to be correct, it must be concluded that the true surface area of these polished quartz crystals is 2.5 times the geometric area.

The $\Delta \nu$ versus P/P^0 curve for water is given in Fig. 5. Again, if a minimal figure of $10 \, A^2$ is assumed for the surface area of a water molecule, it is necessary to accept a roughness factor of 2–2.5 to interpret the data reasonably in terms of Sauerbrey's formulae. The salient feature of the water "isotherm" is the extremely steep initial rise, approximately one quarter of a monolayer being adsorbed at a relative pressure of 10^{-4}.

The change in resonant frequency of a 100-kc X-cut

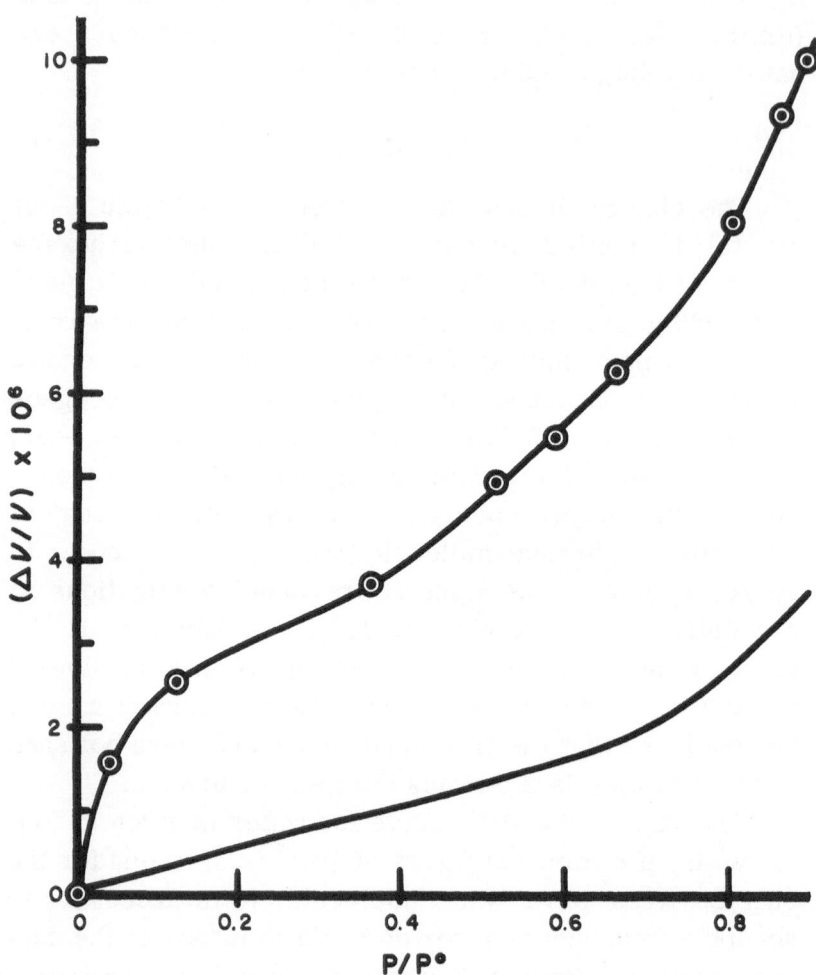

Fig. 4. The change in frequency of a 10–Mc Y–cut crystal versus relative pressure of hexane. $T = 25°C$. The lower solid curve represents an adsorption isotherm for hexane on a 0.9 m²/g quartz powder.

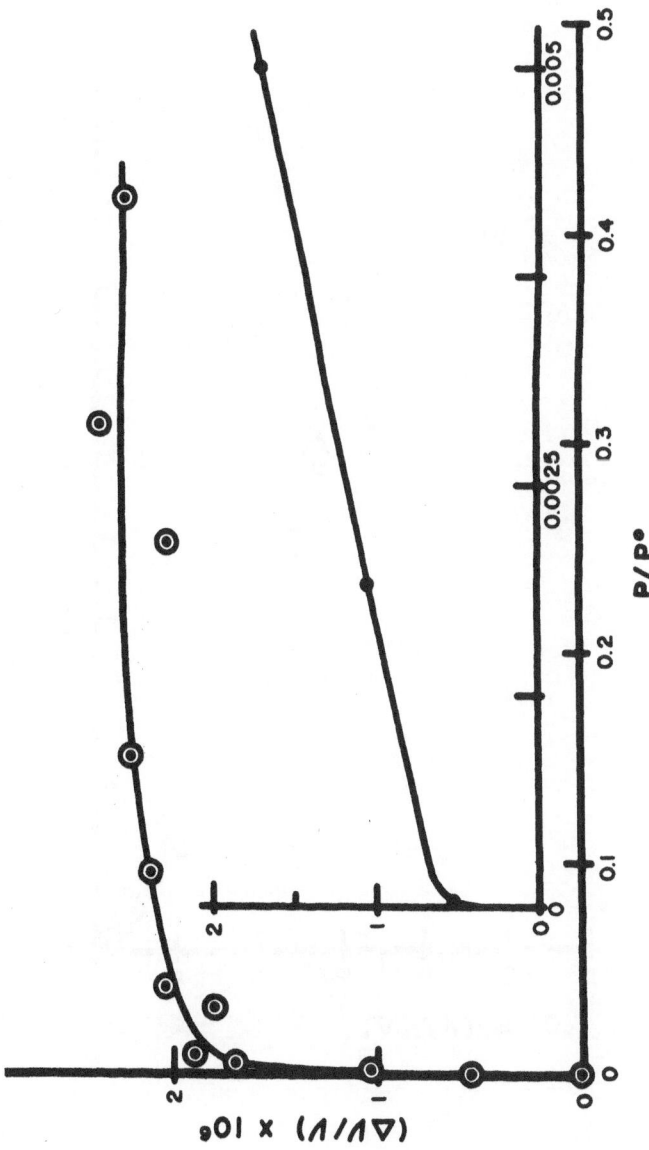

Fig. 5. The change in frequency of a 10-Mc Y-cut crystal versus relative pressure of water at 25°C. The inset figure displays the data at low relative pressures on an expanded scale.

W. H. Wade and L. J. Slutsky

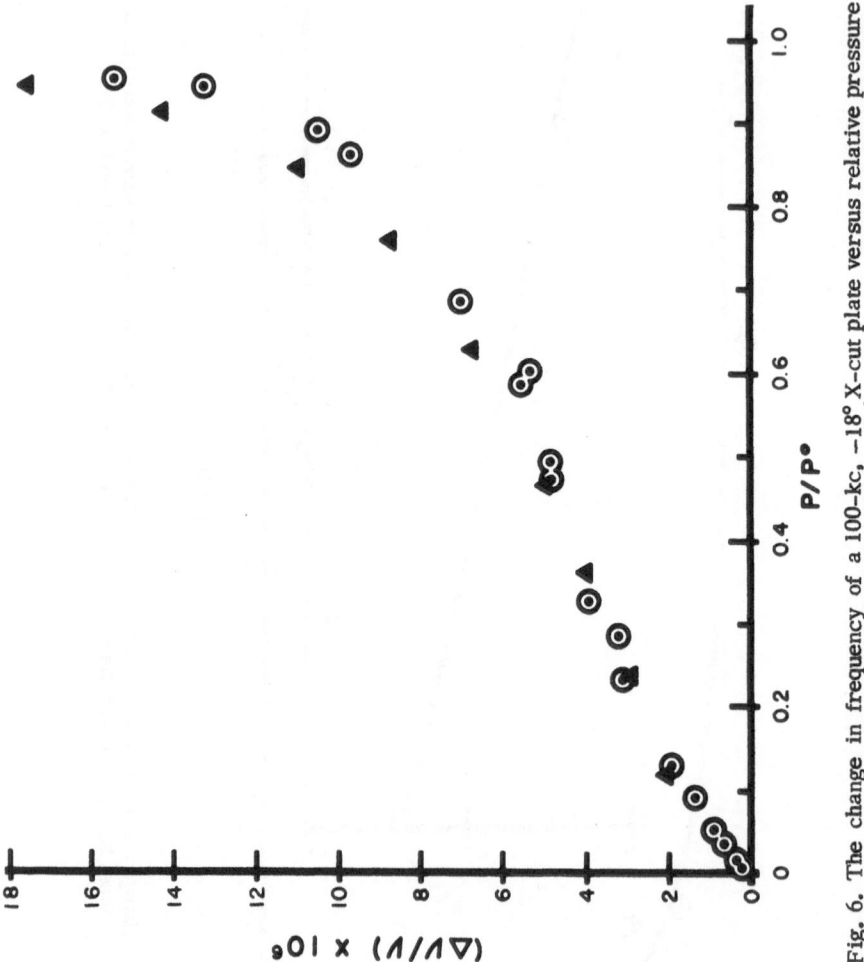

Fig. 6. The change in frequency of a 100-kc, $-18°$ X-cut plate versus relative pressure of hexane. The circles indicate points at $0°C$ ($P^0 = 4.531$ cm) and the triangles at $25°C$ ($P^0 = 14.93$).

plate is plotted versus the relative pressure of hexane in Fig. 6, where the circles indicate points obtained at $0°$ C (P^0 = 4.531 cm) and the triangles points obtained at $25°$ C (P^0 = 14.93). The frequency changes for the X-cut plate are larger than those experienced by the Y-cut crystal, in spite of the greater thickness of the X-cut. Both the X- and Y-cut crystals were polished by the same method, and there should be no substantial difference between their roughness factors. It is thus unlikely that the large frequency changes may be attributed to simple loading of the crystal by the adsorbed mass. However, the similarity in the $\Delta\nu$ versus P/P^0 plots for $0°$ C and $25°$ C strongly suggest that $\Delta\nu$ is proportional to the number of moles adsorbed. Tests with argon indicate that the oscillator frequency is not greatly affected by a nonadsorbing gas at densities and acoustic impedances comparable with those of hexane in the pressure range 0–15 cm.

In the case of an X-cut crystal the elastic strain at the principal faces is not zero, and any alteration of the elastic constants in the interfacial region might be supposed to affect the resonant frequency. In any case, it may be concluded that the change in resonant frequency of crystals vibrating in extensional modes is not simply related to the adsorbed mass by $\Delta m/m = \Delta\nu/\nu$, whereas the results for a Y-cut crystal may be consistently interpreted by means of this formula.

REFERENCES

1. Gunter Sauerbrey, Z. f. Physik 155, 206 (1959).
2. R. A. Heising, Editor, Quartz Crystals for Electrical

Circuits, D. Van Nostrand Company, Inc., New York, 1946.

3. P. Vigoreux and C. F. Booth, Quartz Vibrators and Their Applications, H. M. Stationery Office, London, 1950.

4. N. Hackerman, R. Every, and W. H. Wade, J. Phys. Chem. 65, 25 (1961).

INVAR BEAM BALANCE FOR THE STUDY OF FAST CHEMICAL REACTIONS

Earl A. Gulbransen and K. F. Andrew
Chemistry Department,
Westinghouse Research Laboratories
Pittsburgh, Pennsylvania

ABSTRACT

For the study of fast oxidation reactions at temperatures of 1200 to 1500° C, special balances and associated equipment must be used. A simple Invar beam balance 14.5 cm long and weighing 46 g was constructed using 2- and 3-mil tungsten wire supports to replace our conventional microbalance. The balance has a sensitivity of 0.4 to $0.8 \cdot 10^{-4}$ g/div. The balance is read in the conventional manner at the end supports and 1 division equals 0.001 cm. With this balance weight changes of 0.1 g can be measured on 10-g specimens with a precision of 1 to $2 \cdot 10^{-5}$ g.

Furnace tubes, specimen supports, and furnaces for operating in the temperature range of 1200 to 1500° C are described. Typical oxidation curves are presented.

INTRODUCTION

This paper describes a high-temperature reaction system and a low-sensitivity Invar balance for the study of the kinetics of fast gas–metal reactions. The system was specifically designed for the study of the kinetics of oxidation of tungsten.

Six major problems must be considered in the design of the reaction system, balance, and specimens: (1) To avoid localized reaction at edges smooth cylinders of metal must be used for samples. (2) The balance must be designed for specimen weights up to 10 g. (3) The balance sensitivity should be sufficiently low to allow a wide range of weight changes on a 10-g specimen. (4) Support wires and buckets and the furnace tube must be chosen to avoid vaporization, reaction with oxygen, and reaction with the oxidized specimens. (5) The furnace tube must be gastight and must not be sensitive to thermal shock. (6) The temperature of the specimen must be accurately measured and controlled.

This paper will describe one approach to the solution of the many problems found in working with high-temperature reaction systems. Results will be presented. Areas for improvement will be suggested.

LITERATURE

Microbalance techniques for moderate- and high-temperature applications have been discussed earlier by Gulbransen,[1] Gulbransen and Andrew,[2] and Walker.[3] The application of gastight mullite furnace tubes for high-vacuum high-temperature applications was studied by Gulbransen and Andrew.[4] In a recent paper Walker[3] has attempted to apply microbalance techniques to the study of solid–gas reactions at temperatures up to 2000° C or higher. High temperatures were obtained by induction heating. Blackburn[5] in our laboratories has also used induction heating methods for microbalance studies at high temperature.

EXPERIMENTAL
Size and Shapes of Specimens

Many metals in the form of sheets or ribbons have a laminar structure. This may lead, at high temperature, to excessive reaction at the specimen edges. In addition, heat liberated during reaction at the edges is not removed as fast as from other areas on the sheet specimen. This leads to an increase in temperature at the edges and an acceleration of the rate of reaction. A typical example of the edge type of reaction is shown in Fig. 1. Here tungsten is reacted with oxygen at 950° C and 0.1 atm pressure of oxygen. The nonuniform type of chemical reaction makes the interpretation of the kinetic results very difficult.

For the study of the kinetics of oxidation of tungsten and the evaporation of tungsten trioxide we use two specimen supporting systems. For temperatures up to 1200° C the specimens are tungsten cylinders, 0.317 cm in diameter, having hemispherical ends and various lengths, as shown in Fig. 2B. The specimen weight for the 1.55-cm-long specimen is 2.25 g. The specimen is supported in an alumina or mullite tube lined with a thin platinum tube. Figure 3 shows the arrangement of the sample holder in the reaction and furnace system. The mullite tube and platinum liner have small holes incorporated in the structure to give better access of oxygen. The weight of the specimen supporting system is 1.84 g.

In the second system the specimen is designed so

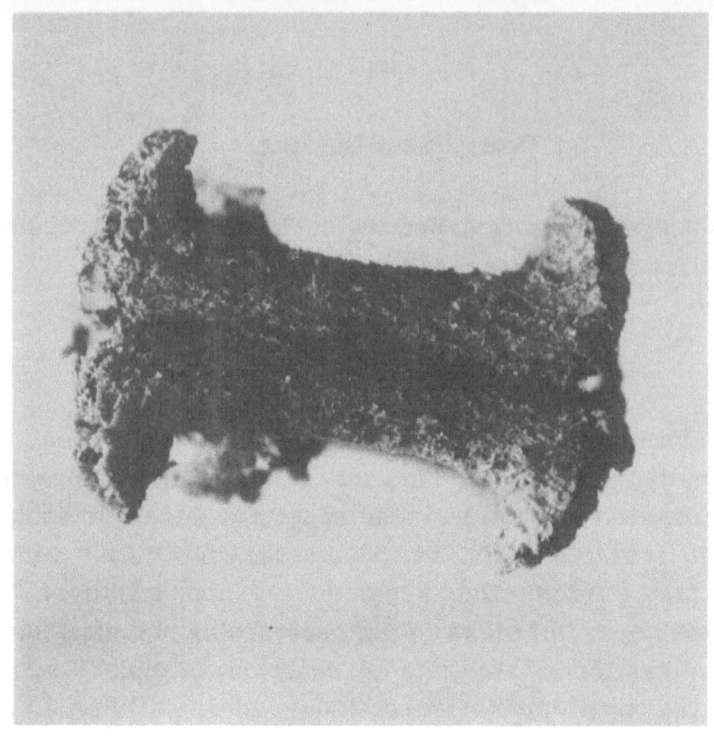

(A)

Fig. 1. Edge reaction of tungsten. Oxidation at 950°C and 0.1 atm of O_2, 54 min, hot-sheared. (A) Edge; (B) surface (20×).

(B)

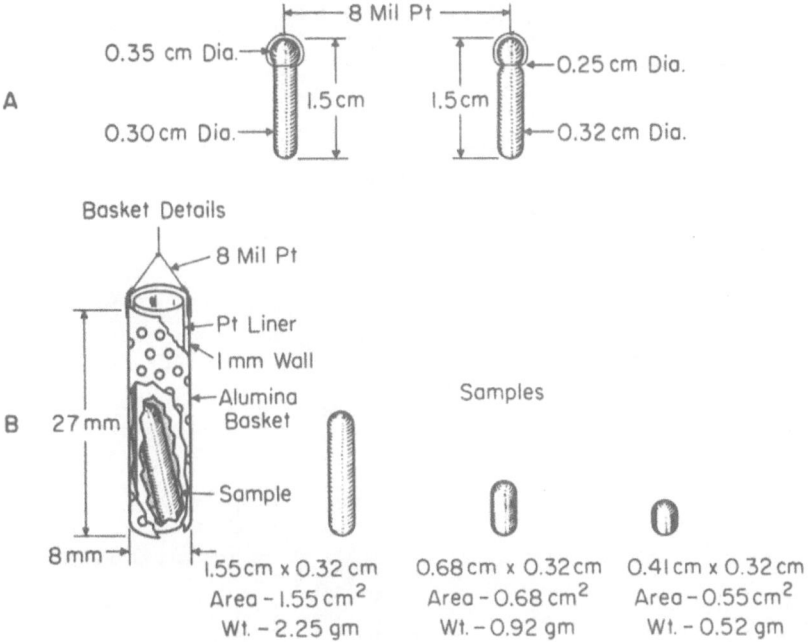

Fig. 2. Specimen support systems and specimens. (A) Direct support; (B) basket support.

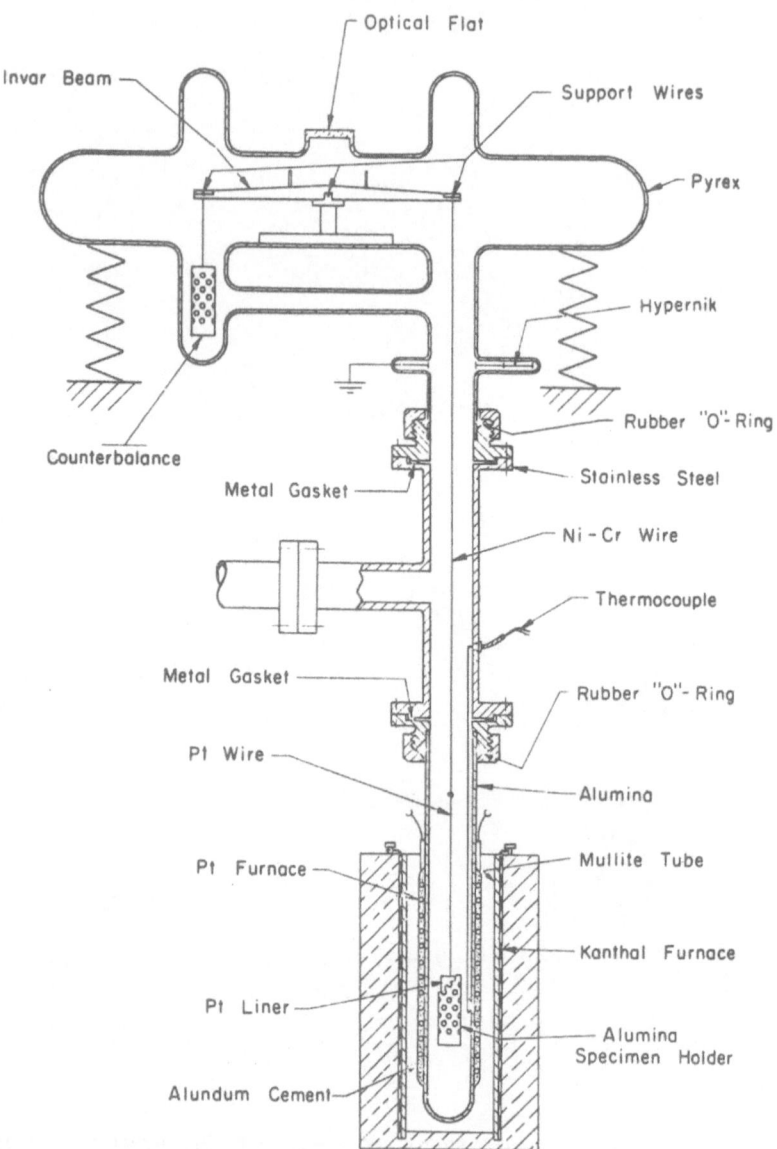

Fig. 3. Microbalance system.

that an 8-mil platinum wire can be used for support. Figure 2A shows two possible specimen shapes which eliminate the edge type of reaction and which also allow a simple support mechanism.

Cylindrical specimen shapes minimize the surface-area-to-weight ratio. This would be a disadvantage in most microbalance work; however, for the study of rapid reactions at high temperatures, the low surface-area-to-weight ratio is an advantage. Thus, in the oxidation of tungsten, where a volatile oxide is formed, it is an advantage to react 1 to 5% of the specimen weight to obtain the proper experimental data for a rate analysis. If we use a 2-g specimen, the change in weight of the sample may be as high as 0.1 g.

Invar Beam Balance

The balance was designed to fit our conventional balance housing. This involves having the balance fit in a 38-mm-diameter glass tube and having a beam length of 14.5 cm (Fig. 3).

The sensitivity of the conventional coplanar gravity type of microbalance is inversely proportional to the beam weight and to the distance of the center of mass to the center of support. For studies on the oxidation of tungsten and evaporation of tungsten trioxide we use specimens weighing about 2.24 g and a basket support weight of 1.84 g. The weight change is 0.1 g. These considerations require a balance sensitivity of 40 to 80 μg per 0.001 cm deflection at 7.25 cm.

Invar was chosen as the construction material because of its high density and its low temperature coefficient of expansion. Figure 4 shows a photograph of the beam and supporting system. The beam is constructed in three pieces. The two end U sections are grooved to fit the central section and then fastened together by Invar screws. To make final adjustments two adjusting weights are used at each end of the balance beam. Two Invar weights located near the central support section are used to adjust the center of mass of the balance relative to the center of support. The support system is also made of Invar. All parts are gold plated to prevent rusting. The total weight of the beam is about 46 g.

The central support wire for the beam is a 0.0076-cm annealed tungsten wire, and the end wires are either 0.0051- or 0.0076-cm annealed tungsten wires. Small grooves are made in the Invar metal for locating the support wires. Invar plates and screws are used for mounting the wires to the beam and supporting frame.

Since the end specimen supports and the central beam support are coplanar, the balance sensitivity should be independent of the load. In spite of the fact that a weight of 640 g was used for stretching the wires during mounting on the frame, some sagging occurs in the mounting wires due to the weight of the beam and the specimens. This sagging lowers the sensitivity as a function of the load.

Calculations show that a sensitivity of $4.7 \cdot 10^{-5}$ g for a 0.001-cm deflection at 7.25 cm should be found using a 46-g beam and a distance of 0.0545 cm from the center of gravity to the center of support. The actual sensitivity was $4.1 \cdot 10^{-5}$ g per 0.001 cm deflection using 0.0076-cm tungsten support wires both on the beam and on the end sections. A period of 2 to 3 sec was found.

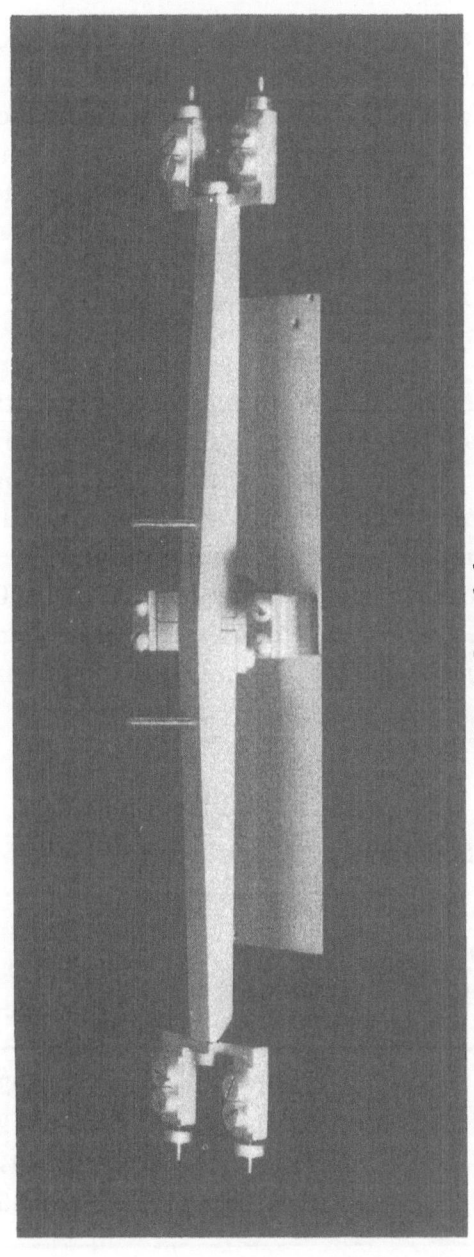

Fig. 4. Invar balance.

Table I shows the effect of load on the sensitivity of the Invar balance.

In the design of the tungsten wire support calculations show that the torsion due to the twisting of the tungsten wire is less than one-tenth of the restoring force due to gravity.

Specimen Support Systems

Several factors must be considered in choosing a material for the specimen support. First, the material must have a high melting point, a low vapor pressure, and a low rate of reaction with oxygen. Second, the material must not react with the specimen or with the oxides formed in the specimen. Platinum meets the vapor pressure requirement up to a temperature of 1600° C. Mullite and alumina react with tungsten trioxide when in contact. To avoid this reaction the mullite or alumina buckets are lined with platinum.

TABLE I

Sensitivity vs. Specimen Weights

Weight on beam, g	Weight added, g	Sensitivity, g per 0.001 cm deflection
0.0	0.0182	$0.4083 \cdot 10^{-4}$
0.9672	0.0182	$0.6212 \cdot 10^{-4}$
3.0017	0.0182	$0.7794 \cdot 10^{-4}$
5.2256	0.0182	$0.8465 \cdot 10^{-4}$

Figure 3 shows the balance assembly, specimen support, and furnace system. For fast reactions where spalling of the oxide occurs, the bucket type of support system allows one to collect the oxide lost by spalling. In an earlier system, alumina or mullite buckets were used alone. Unfortunately, reaction occurred above 1200° C between the ceramic and tungsten trioxide. The platinum liner minimizes this reaction. This system works for temperatures up to 1200° C, where spalling of the oxide is sometimes observed. A 0.2-mm platinum wire is used to support the bucket in the hot zone of the furnace. A 0.005-cm Nichrome wire is used to support the system in the colder part of the furnace system.

When direct-supported specimens are used, the furnace tube of Fig. 3 is extended 2 in. lower in the furnace box. This allows collection of spalled reaction products in a cooler part of the furnace system. Tungsten trioxide spalling into the furnace tube at 1200° C and higher would form a mixed oxide with the alumina. It is necessary, therefore, to cool the bottom of the furnace tube. Again a 0.2-mm platinum wire is used for supporting the tungsten in the hot zone of the furnace.

Furnace Tubes

Gastight mullite and alumina furnace tubes are used. These were obtained from the McDanel Refractory Company. Since the refractory materials react with tungsten trioxide vapors the life of a furnace tube is limited. Easy dismounting of the tube is essential for replacement of the tube as well as collection of spalled oxide. A rubber O-ring connection is used for connection to the ceramic or to the glass balance housing. Air is passed over the connection for cooling. Metal gaskets are used to connect the metal parts of the system.

Table II shows the vacuums achieved in the reaction system at room temperature and at 1000° C. Both McLeod gauge and Bayard-Alpert ion gauge methods were used. A typical leak rate measurement is also shown in Table II.

Both the mullite and alumina furnace tubes were insensitive to the thermal shock produced by a rapid raising or lowering of the furnace around the furnace tube.

TABLE II

Vacuum Readings of Mullite Tube Furnace
and Leak Rates

Temperature	Elapsed time, min	McLeod gauge, mm Hg	Ion gauge, mm Hg
R.T.*	–	$1 \cdot 10^{-6}$	$7 \cdot 10^{-7}$
1000° C	–	$5 \cdot 10^{-6}$	$3 \cdot 10^{-6}$
R.T.	0	$1 \cdot 10^{-6}$	$1 \cdot 10^{-6}$
R.T.	10	$1 \cdot 10^{-5}$	$1 \cdot 10^{-5}$
1000° C	0	$5 \cdot 10^{-6}$	$3 \cdot 10^{-6}$
1000° C	10	$1 \cdot 10^{-5}$	$2 \cdot 10^{-5}$

*Room temperature 25°C.

Furnaces

The furnace has over-all dimensions of 6.5 in. width, 6.5 in. length, and $12\frac{3}{4}$ in. height. The outside walls are made of transite. Two furnace elements are used. For temperatures up to 1200° C a Kanthal strip furnace winding is used. This furnace unit is wound on a $2\frac{1}{4}$-in.-diameter, $\frac{1}{8}$-in.-wall mullite tube and cemented with Norton RA1139 alundum cement. The resistance wire is $\frac{1}{4}$-in.-wide, 10-mil-thick Kanthal A strip. Silicell is used as insulating material. For temperatures above 1200° C a platinum winding cemented to the alumina furnace tube is used to add to the heat supplied by the Kanthal furnace unit. The platinum heater is 40 mils wide and 6 mils thick and is cemented to the alumina furnace tube by alundum cement. Table III shows the power requirements for the Kanthal furnace operating alone at 1200° C and for the combined furnace system operating at 1450° C. This furnace system is compact, easy to maintain, and easy to control to a temperature of 1500° C.

Temperature Measurement

A platinum—platinum-10%-rhodium thermocouple is used. To assure that the temperature of the specimen is measured, the thermocouple is mounted inside of the furnace tube adjacent to the specimen. The leads are brought out of the reaction system by means of a metal—glass seal, as shown in Fig. 3. The thermocouple is mounted in a two-hole alundum protection tube. The thermocouple is calibrated against a National Bureau of

TABLE III

Power Requirements of Furnace

Resistance furnace	Operating temper- ature, °C	Current, amp	Voltage, v	Power, w	Total power, w
Kanthal	1200	11.5	55.0	632.5	632.5
Kanthal*	1450	11.8	55.0	649 ⎫	804
Platinum*	1450	2.5	62.0	155 ⎭	

*Kanthal and platinum furnaces operating as a single unit.

Standards standard, and the recorder-controller is calibrated against a standard voltage box.

RESULTS

Figure 5 shows a typical curve for the reaction of a direct-supported tungsten sample with oxygen at 3.8 cm Hg of O_2 at 1200° C. The weight gain in milligrams per square centimeter is plotted against the time in minutes. The sample weight was 0.8385 g, and the sample area was 0.693 cm^2. Figure 5 shows that the tungsten trioxide which forms is quickly volatilized. After the first 7 min of reaction the weight loss followed a linear law. A rate of 0.0866 mg/cm^2-sec was calculated. After 24 min of reaction 13% of the mass of the tungsten had reacted and volatilized.

The balance used in this experiment has a sensitivity of 76 μg per 0.001 cm deflection at 7.25 cm.

DISCUSSION

The results of experiments on the oxidation of tungsten show the low-sensitivity balance under discussion to be a very useful instrument in following fast chemical reactions at temperatures in the range of 1200 to 1500° C. A number of difficulties still remain. (1) A better type of ceramic material which will not react with WO_3 solid or $(WO_3)_3$ vapor at temperatures of 1200° C and higher must be found for the furnace tube and sample support. This reaction causes discoloration of the ceramic tube and eventually leads to cracking of the tube. In our present furnace arrangement this involves a reassembly of the platinum furnace winding on a new refractory tube.

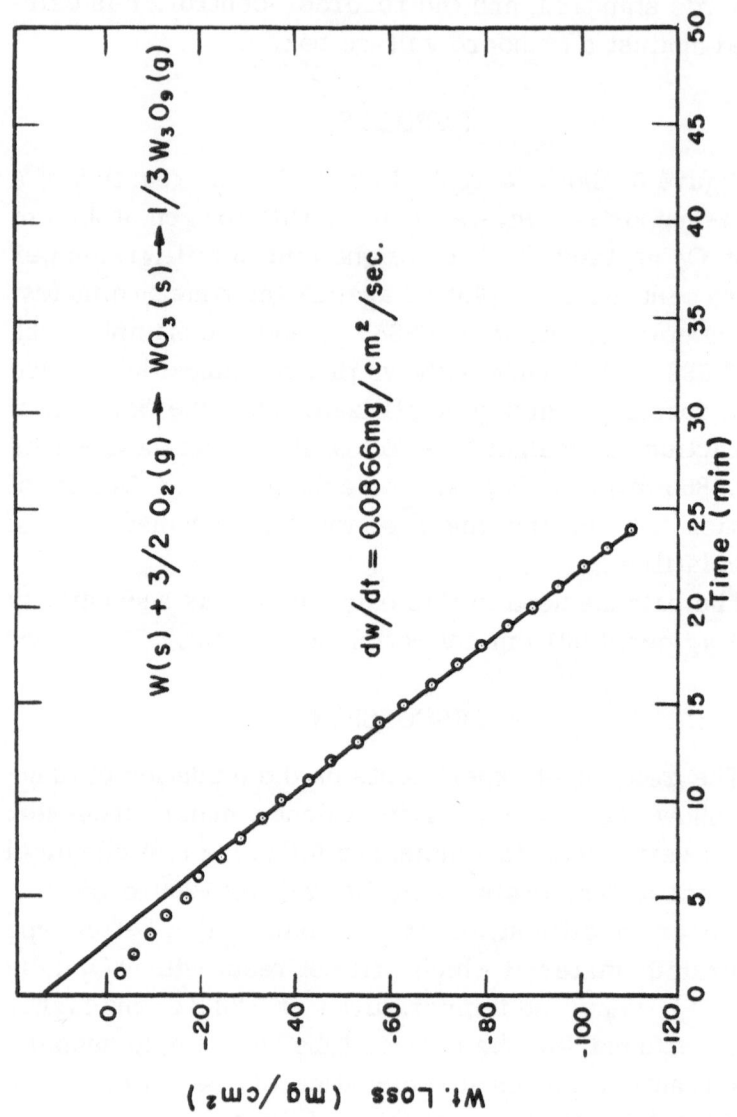

Fig. 5. Oxidation of tungsten at 1200°C and 3.8 cm Hg of O_2.

(2) Since WO_3 is absorbed in the furnace tube, condensation of WO_3 solid occurs on the support wires. This effect can be minimized by heating the tube to the reaction temperature in pure argon. (3) Weight gain measurements alone cannot give the necessary kinetic data. Oxygen reacts with the sample while tungsten trioxide is evaporating from the surface. A separate measurement of oxygen uptake is required. We are now working to develop such a reaction system.

REFERENCES

1. E. A. Gulbransen, Advance in Catalysis, Vol. 5, Academic Press, New York, 1953, p. 119.
2. E. A. Gulbransen and K. F. Andrew, Vacuum Microbalance Techniques, Vol. 1, Plenum Press, New York, 1961, p. 1.
3. R. F. Walker, Vacuum Microbalance Techniques, Vol. 1, Plenum Press, New York, 1961, p. 87.
4. E. A. Gulbransen and K. F. Andrew, Ind. Eng. Chem. 41, 2762-2767 (1949).
5. P. E. Blackburn, Communication to 2nd Microbalance Conference.

AN APPARATUS FOR THE SIMULTANEOUS MEASUREMENT OF THE OPTICAL TRANSMISSION AND MASS CHANGES IN THIN FILMS

A.W. Czanderna and Harold Wieder

ABSTRACT

An apparatus which has been employed for the measurement of the optical transmission and mass changes in metal films is described. The stoichiometry of the metal film is inferred from measurements carried out with an automated bakeable vacuum microbalance. The optical transmission of the film is measured for monochromatic light from 4000 to 8000 A. The experimental difficulties encountered in this study are discussed in detail.

Results obtained with this apparatus during the stepwise evaporation of a copper film are presented and briefly discussed.

Some pertinent comments concerning the design and use of pivot-type microbalances are included in an appendix.

INTRODUCTION

In the past two decades careful design and construction of beam microbalances have resulted in an increased application of vacuum microbalance techniques. Recently, descriptions of automatic recording microbalances and balances that can be baked and operated in the 10^{-10} torr pressure region have appeared. These developments represent advanced stages of microbalance design employed for the purpose of measuring one parameter, namely, mass change. The next natural extension of microbalance techniques is the integration of the mass measurement with measurements of other parameters. The dynamic nature of micromass changes requires that the additional parameters be measured simultaneously. Some of these parameters are magnetic susceptibility, spin resonance, electrical conductivity, thermoelectric power, optical absorption, and ambient gas composition. Clearly, some of these measurements would involve far greater experimental difficulty than others. For example, the measurement of thermoelectric power would require that several lead wires be attached to a suspension which must also swing freely, whereas magnetic susceptibility and gas analysis apparatus need not make mechanical contact with the balance.

Magnetic susceptibility measurements have been carried out utilizing a beam microbalance, but not as an additional parameter.[1] A bakeable balance with an omegatron for analysis of the ambient gas has also been described.[2]

In this paper a technique for simultaneously measuring the mass change and optical transmission of supported thin films will be described. To date measurements have been obtained during the oxidation of copper films ranging in thickness from 200 to 1500 A.[3] However, it seems likely that information of considerable value could be obtained on other films utilizing this technique.

For carrying out this study, a thin metal film is evaporated onto a transparent substrate suspended from a microbalance. During subsequent oxidation of the film, the mass of oxygen incorporated and the optical transmission of the film are continuously and simultaneously measured. The mass measurements characterize the stoichiometry; the optical measurements provide fundamental physical information about the film and a "fingerprint" of the stoichiometry.

The integrated experimental apparatus will be described separately under mass measurements, film preparation, and optical measurements.

MASS MEASUREMENTS

Mass changes are determined with a pivot-type beam microbalance. Details of the construction, operation, calibration, and limitation of the pivot-type balance have already been described.[4,5] The balance was modified to permit bakeout at 400° C.* The operating characteristics of the modified balance include a reproducibility of ± 0.1 μg, a period of 12 sec, and a sensibility of ± 0.2 μg in high vacuum and ± 1 μg during oxidation. The change in sensibility arises because of buoyancy and thermomolecular flow effects.[6]

A Pyrex glass plate substrate (A), 0.25 mm thick and 40 mm in diameter and weighing 0.7 g, was suspended from the balance (B), as shown in Fig. 1. A 6-mm-diameter hole was drilled in the field lens (C) to allow

*Details of the modifications are presented in an Appendix at the end of this paper.

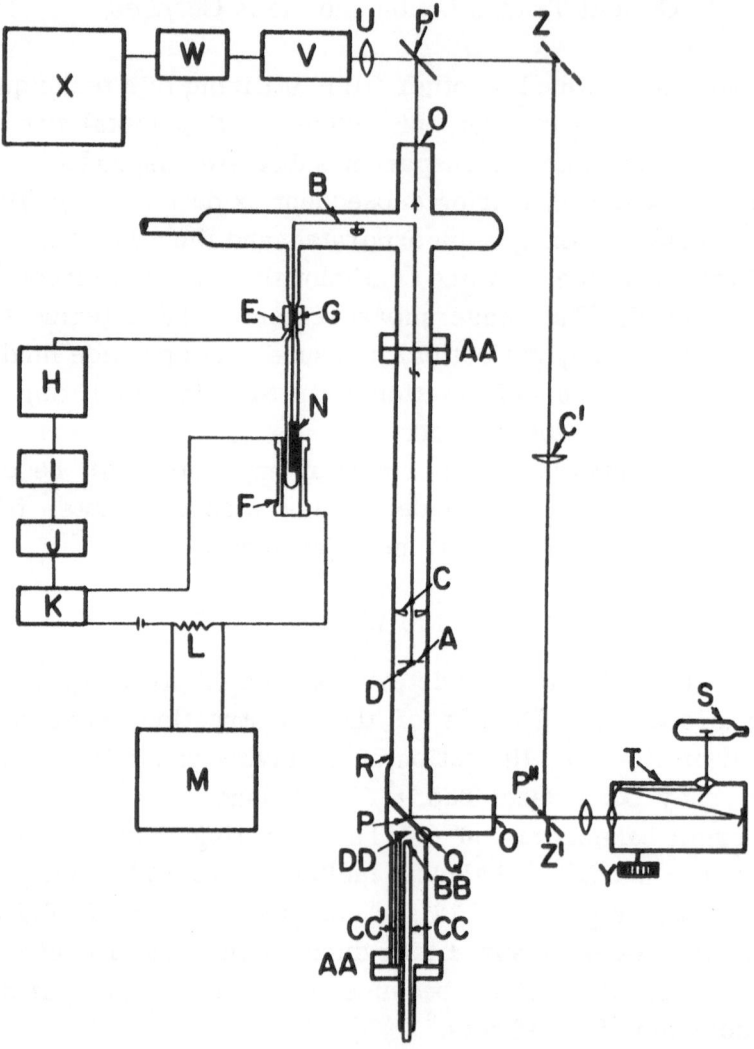

Fig. 1. Apparatus for simultaneous mass and optical transmission measurements. (A) Pyrex glass substrate; (B) beam microbalance; (C, C') field lens; (D) quartz disc; (E) transducer; (F) magnetic compensation solenoid; (G) piano wire probe; (H) translator; (I) amplifier; (J) servomotor; (K) Helipot; (L) standard resistor; (M) bucking voltage sources and recorder; (N) Cunife compensation magnet; (O) quartz windows; (P, P', P") silvered mirrors; (Q) slotted Pyrex glass rod; (R) square Pyrex glass tubing; (S) light source; (T) monochromator; (U) opal glass disc; (V) photomultiplier; (W) photometer; (X) recorder; (Y) motor drive; (Z, Z') mirror positions; (AA) stainless steel flanges; (BB) molybdenum boat; (CC, CC') copper rods; (DD) nickel shield.

adequate clearance of the suspension fiber. The substrate was held in a horizontal position by a quartz disc (D), 6 mm in diameter and 1 mm thick. The disc was cut from 0.5-mm-ID quartz capillary tubing, using a diamond saw. A quartz rod, 0.5 mm in diameter, was passed through the hole of the disc and fused to the bottom. Use of a small flame at the point of fusion prevented warping of the disc.

The requirements demanded by the optical measurements resulted in a suspension design which was of such poor geometry that thermomolecular effects were detected at pressures as low as $3 \cdot 10^{-6}$ torr. This could not be improved with a symmetrical system,[6] so the simpler asymmetrical system was used.

Automatic operation of the balance was achieved utilizing a transducer (E) and magnetic compensation (F, N), as has been described.[7] A 0.485-in. length of piano wire (G), 0.031 in. in diameter, was found to produce the optimum signal change when employed with a JC-2-0.25 transducer and a Model 83 Type F translator (H), manufactured by the Crescent Engineering and Research Company.

In automatic operation the balance is maintained at the null position. Deflection of the balance moves the probe in the transducer and results in unbalancing of an ac bridge circuit formed by the transducer and translator circuits. The unbalanced voltage is amplified and rectified by the translator, producing a positive or negative voltage signal depending on the direction of movement of the balance. The output signal of the translator is amplified (I) and used to drive a servomotor (J). The action of the servomotor on a 1000-ohm 40-turn Model E Helipot (K) results in a change in the current in the solenoid (F), restoring the balance to the null position and changing the voltage drop across a standard resistor (L). The latter voltage is recorded after being reduced to between 0 and 1 mv by two bucking voltage sources (M), one of which is automatically changed in 1-mv steps by travel of the recorder pen to either extreme of a −0.1 to 1.0 mv scale.

In order to attain very high vacuum, it was found necessary to place the transducer outside the vacuum system. A silver tube, 0.125 in. OD and 0.119 in. ID, was used as a liner for the transducer. Silver was chosen because it fulfills the need for (1) a bakeable vacuum enclosure, (2) an electrical ground that eliminates electrostatic interactions between the glass-enclosed probe and the vacuum wall, and (3) an excellent thermal conductor which minimizes the thermal gradients in the stainless steel core of the transducer.

The weight of the Cunife compensation magnet (N) was found sufficient to overcome lateral forces which act on the probe in the transducer core at zero vertical pull. These lateral forces are very small for a translator with a 20-kc frequency. They do not constitute a problem as long as the probe can be weighted.

OPTICAL MEASUREMENTS

The design of the optical system was governed in part by other experimental requirements. For example, the distance between the quartz windows (O) and, hence, the minimum optical path length, is 105 cm. In order to confine the optical beam to 40 mm in diameter, it is necessary to insert a field lens (C, C') at the midpoint of the optical path. The field lens in turn is the principal cause of the excessive thermomolecular flow effects discussed earlier.

With the exception of the field lens (C) and one mirror (P), all optical components are mounted on a rigid bench and are relatively easy to align. Since the field lens is placed in an image plane, its optical tolerances are rather generous. The alignment of the mirror (P) represents a difficult problem. The mirror fits into a slotted Pyrex glass rod (Q), which can be rotated within its enclosure by an attached soft iron core. Alignment is achieved by adjusting the over-all length of the mirror until in its fully rotated position it makes a 45° angle with the direction of the beam. A section of square Pyrex glass tubing (R) provides ample room for the oval-shaped mirror when it is in the vertical position.

In the optical apparatus shown in Fig. 1 monochromatic light is obtained from a ribbon-filament tungsten source (S) which is incident on a Bausch and Lomb grating monochromator (T). The exit slit of the monochromator, which is adjusted for a 10-A bandwidth, is magnified 3 : 1 and imaged onto the field lens (C, C'). The image is then demagnified 1 : 3 and refocused onto an opal glass disc (U) located just in front of an RCA 6217 photomultiplier tube (V). An Aminco photometer (W) is used as the photomultiplier power supply; the photovoltage is displayed on a 10-mv recorder (X). The combination of photometer and two neutral density screens which may be rotated into the beam yields an over-all range of 10^5 in the gain.

The wavelength region from 4000 to 8000 A is scanned by means of a motor drive (Y) attached to the wavelength drum of the monochromator. After the sample path is scanned, the upper mirror (P') is moved to position (Z) and a mirror is inserted at position (Z'). The equivalent blank path is then scanned. The transmission of the sample at any wavelength then is the ratio of the photovoltage obtained for the sample path to that obtained for the blank path. Reference markings at 1000-A intervals on the chart enable an accurate wavelength comparison to be made.

To avoid scattering from the hole in the field lens the center portion of the slit is masked. Scattering from the microbalance beam and suspension cannot be avoided, however. The magnitude of this scattering is determined in a run made before each evaporation and is used to correct all subsequent transmission measurements for that particular film.

VACUUM TECHNIQUE AND FILM PREPARATION

Vacuum in the system is produced with a forepump, a three-stage oil diffusion pump, and a zeolite trap. The limiting pressure of $5 \cdot 10^{-9}$ torr attained after bakeout presumably results from the use of Teflon plastic gaskets in the stainless steel bellows-type valves utilized in the gas-handling system. Gold gaskets are used to attain vacuumtight seals at the stainless steel flanges (AA). Heating tapes are used to bake out the system instead of a furnace because of the geometry requirements of the optical bench. The pressure in the vacuum system is read with a Bayard-Alpert ionization gauge or a mercury manometer protected with a trap.

For deposition of a film on the substrate, the silver mirror (P) enclosed in the vacuum system is rotated to a vertical position. The evaporator assembly consists of a molybdenum boat (BB) for heating the charge, supported by two $\frac{1}{4}$-in.-diameter copper rods (CC). Large copper rods were found necessary to concentrate the heat at the molybdenum boat, and thus minimize the outgassing problem. The vacuum seal for the copper rods was prepared by a three-step process. Each rod was first silver-soldered to a cylindrical stainless steel ring at the high-vacuum end. After the soldered joint was thoroughly cleaned, the ring was welded first to one end of $\frac{3}{4}$-in.-diameter double Kovar joint and then to the stainless steel flange; this was performed by the Heliarc welding process. This procedure insured that any flux trapped between the copper rod and the stainless steel ring was not exposed to the vacuum system. No vacuum difficulties have been experienced with this assembly. A nickel shield (DD), which is supported from a third copper rod (CC'), is used to cover or expose the molten charge. The 25-cm distance from the boat to the substrate produces a theoretical film uniformity of about 1%. During bakeout of the vacuum system, the charge is degassed several times by heating to a point slightly below the evaporation temperature. After the vacuum system is baked out and cooled, several small bursts of the evaporant are deposited with the charge covered. This deposition serves to purify the charge further, and in the case of a metal evaporant, provides a gettering surface near the substrate. After deposition of the getter, and when a pressure of 0.1 μtorr or better is achieved, the charge is exposed and heated.* Onset of deposition is observed as an apparent mass loss resulting from the

---

*In the design discussed here, pressure measurements were made close to the pump rather than near the evaporation source as was done in a prototype system. However, experience from the previous system showed that the pressure at the evaporator did not differ by more than a factor of two from the pressure at the pump during evaporation.

momentum transfer of the impinging evaporant. Deposition is allowed to occur until a pressure not greater than 1 μtorr is reached. After a pressure of 0.1 μtorr or better is again attained by pumping and gettering, the evaporation is resumed.

In the preparation of a typical 500-A-thick copper film, several evaporation intervals ranging from 30 sec to 4 min each were employed. The total deposition time was about 10 min and the integral of $P\,dt$ was usually about 2 μtorr minutes. The amount of copper deposited was automatically recorded with the balance during each pump-down interval. After the desired film thickness was reached, the mirror was rotated to the 45° position for optical transmission measurements.

RESULTS AND DISCUSSION

The oxidation and reduction of copper films using a prototype system have been discussed elsewhere.[3,8] As an example of the increased versatility of the present apparatus, the transmission of a copper film of increasing thickness is shown in Fig. 2. The film was evaporated to a desired "thickness," which was determined from the mass gain. The mirror was rotated and the transmission of the film was measured. The mirror was then rotated out of the path of the evaporating source, and a second evaporation was made. This process was repeated ten times until a final film thickness of approximately 500 A was reached. The transmission curves obtained after each evaporation step can be seen in Fig. 2. This experiment, while quite simple, is nevertheless very revealing. For example, a plot of transmission vs. thickness for any

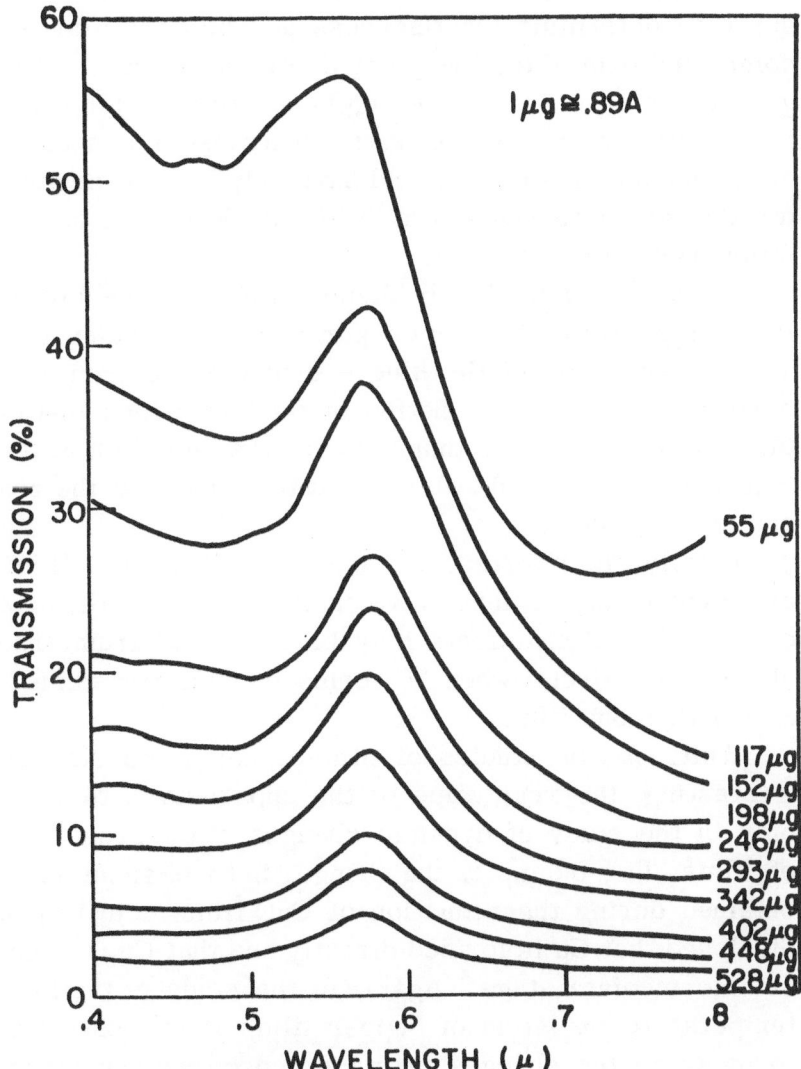

Fig. 2. Optical transmission spectrum of a copper film at various stages of evaporation.

wavelength can be compared with that of a theoretical curve to determine the thickness at which the film becomes uniform (i.e., the point at which the copper aggregates merge). Since the agglomeration of thin films is a function of evaporation rate, substrate temperature,* etc., such a comparison would have only limited meaning for the set of curves shown in Fig. 2. Hence, the calculation was not carried out.

A second point clearly demonstrated by the curves is the high reflectivity of copper at longer wavelengths. The transmission of the blue end of the spectrum is a much more sensitive function of thickness than that of the red end, showing that most of the light lost at the blue end is absorbed, while most of that lost at the red end is reflected.

Finally, the interference effects usually found in films are evident at the long-wavelength end of the thinnest copper film. Such effects may be separated from true absorption effects when a series of films is studied, rather than one film.

While *in situ* studies of films of known density are interesting, the real value of the apparatus is realized only in the study of dynamic changes of the properties of films. For example, the changes in the transmission obtained during the reduction of CuO films in hydrogen disclosed that Cu is formed directly and that Cu_2O is not an intermediate stage.[8] Again, in the study of the low-temperature oxidation of copper films it was essential to measure the changes in mass and optical spectrum simultaneously in order to characterize the composition $CuO_{0.67}$.[3] Thus it is evident that the simultaneous meas-

*See, for example, Ref. 9.

urement of the optical transmission in itself adds considerable leverage to the microbalance data.

In addition to the above advantages, repeated measurements on copper films have demonstrated that for a given film thickness, the optical transmission bears a unique relationship to the amount of oxygen incorporated into the film. The original film thickness can be controlled during deposition in auxiliary systems by transmission methods. Films of various but known oxygen content can then be prepared for other studies such as conductivity, diffraction, susceptibility, and resonance. Thus, the transmission technique is used to "fingerprint" a dynamic process and allow certain parameters to be studied with all the convenience of a static process.

CONCLUSIONS

Limitations of the Apparatus

Studies with this type of apparatus are limited to films. The thickness region for the study of the oxidation of copper is 200–1500 A. Below 200 A the precision to which changes in mass can be monitored approaches several percent, while for thick films the transmission becomes insensitive. The poor geometry of the system restricts the useful ambient pressure region to 100–400 torr (other than high vacuum) because of thermomolecular flow forces and convection currents. However, the ambient pressure could be composed of gases with various partial pressures, which would effectively extend the pressure range. Radiation from the furnace becomes detectable in the transmission curves at temperatures exceeding 400° C, although light-chopping techniques can be used to avoid this. Extension of the temperature region to below 25° C presents increasingly difficult design problems.

Future Studies

While the above restrictions appear quite limiting, a large number of worthwhile experiments can be performed. Examples of useful studies include the effect of pressure and temperature on the oxidation and reduction of films, and the influence of the substrate material, deposition rate, substrate temperature, and film annealing temperature on the properties of the films.

Modification of the apparatus could be made to include reflection measurements, and the optical range could be extended to the ultraviolet and near infrared regions. The transmission curves can be used in principle to calculate the optical bandgap and its temperature dependence for any stable or metastable composition formed in films.

ACKNOWLEDGMENTS

The authors are pleased to acknowledge the patience and helpful suggestions of D. Shipley and H. Roberts in constructing the microbalance, housing, and hangdown tubes. The authors are also grateful to C. N. Cochran and J. M. Honig for releasing unpublished information which proved helpful in constructing this apparatus.

Appendix

Modification of a Pivot-Type Beam Microbalance
for Use in a Bakeable Vacuum System

For a bakeable system it seemed essential to eliminate all silver chloride seals on the microbalance. The following modifications of the pivot-type balance described previously[4,5] have been found to provide satisfactory operation.

The balance beam is constructed completely from 1-mm quartz rod, utilizing a graphite jig. The quartz cups were fused to the beam, eliminating silver chloride seals for the tungsten points. The 0.003-in. tungsten crosswires at the "Y" were replaced with 1-mm quartz rod that was drawn to a 0.003-in. quartz thread over a 1-mm length. The drawing operation could be conveniently carried out by first aligning and fusing a slightly necked quartz rod onto the beam in the graphite jig. After removing the beam from the jig, concentration of a tiny microflame at the exact center of the necked portion results in drawing of a fiber by surface tension. The drawing process was viewed with a 40× measuring microscope and allowed to continue until the desired cross-fiber size was achieved. The contour of the fiber near the minimum dimension is roughly parabolic. Lateral motion of the 0.003-in. tungsten suspension hook is not possible.

Electropolished 0.020-in. tungsten points were friction-fitted into supports on the balance support frame. For this it was found convenient to fuse a 15-mm length of quartz tubing, 1.5 mm OD, 0.5 mm ID, onto a quartz rod for aligning and supporting the tungsten points. The points and quartz tubing lengths were carefully matched before final assembly. This is extremely critical since any motion of the tungsten points will result in mispositioning of the points in the quartz cups on the beam (see Ref. 4).

Since the transducer probe on one hangdown suspension must hang freely inside a 0.119-in.-ID tube and the substrate suspension plate fiber must clear the 0.250-in. hole in the field lens, the quartz balance frame should be positioned exactly in the balance housing. This positioning requirement was satisfied by machining a graphite "half-moon" with four 6-mm-long graphite protrusions located at dimensions corresponding to those of the "feet" on the balance frame. Positioning of the balance frame in the housing was achieved by first aligning 18-in. lengths of rod that were screwed into the graphite half-moon at the appropriate center-to-center distance and then heating the balance housing until four indentations were produced by the graphite protrusions. Thus, the ends of the balance beam are not only precisely positioned over the center lines of the hangdown tubes, but the frame is also situated in a horizontal plane.

In Ref. 7 several questions were raised concerning design limitations or improvements of pivot-type balances. The modifications described above have answered the question of inversion of the pivot system. Concerning the long-term stability of a calibration factor, it has been found that the same calibration constant and precision have been in effect for a pivot-type balance for three years without changing the tungsten points; the beam plus sample mass is 5 g. Changes in the center of gravity resulting from plastic flow of the points should not be detected as a change in the calibration, which is fixed by the geometry of the compensation magnet and the solenoid, but rather as a change in the precision of the balance.

It seems that exercising due care in selecting tungsten wire that is hard drawn and extreme care to avoid annealing the tungsten during mounting in the balance beam or frame is the essential point.

Special Comments about Automatic Operation of a Pivot-Type Microbalance

It has been found that the size of zero shifts resulting when the beam strikes the arrest can be limited to about ± 2 μg, if the free swing of the beam is limited to 1 mm or less. Larger zero shifts do occur if the balance is subjected to a lateral impact. However, zero shifts of the latter type do not occur except through carelessness. Rapid flow of gases during changes in pressure in the vacuum system or rapid deposition rates for films can produce large excursions of the beam. Under these extreme dynamic conditions, even the most practiced operator can occasionally have data ruined by a zero shift from a freely swinging beam banging the arrest. These operational zero shifts are eliminated by utilization of the transducer null technique for automatic operation because excursions resulting from gas pressure or deposition forces of the order of several dynes are confined to the free swing limits of the beam. This results because motion of the piano wire probe in the alternating magnetic field of the transducer is always opposed by induced magnetic fields in the probe on the balance beam.

The stabilization of the beam by the transducer and probe is a function of the probe diameter and length and of the frequency of the excitation of the transducer. For a 3-kc frequency, a 0.485-in. length of 31-mil piano wire produces the optimum signal to noise level. The diameter was chosen from alignment considerations, and the frequency was fixed by the translator. Zero shifts which may result from lateral impact on the balance housing are recorded as a step in the strip chart recording of the data.

REFERENCES

1. F. Blaha, Mikrochemie ver. Mikrochim. Acta 39, 339 (1952).
2. S. P. Wolsky and E. J. Zdanuk, Vacuum Microbalance Techniques, Vol. 1, Plenum Press, New York, 1961, p. 35.
3. H. Wieder and A. W. Czanderna, J. Phys. Chem., in press, March, 1962.
4. A. W. Czanderna and J. M. Honig, Anal. Chem. 29, 1206 (1957).
5. J. M. Honig, Vacuum Microbalance Techniques, Vol. 1, Plenum Press, New York, 1961, p. 55.
6. A. W. Czanderna, Vacuum Microbalance Techniques, Vol. 1, Plenum Press, New York, 1961, p. 129.
7. C. N. Cochran, Vacuum Microbalance Techniques, Vol. 1, Plenum Press, New York, 1961, p. 23.
8. H. Wieder and A. W. Czanderna, J. Chem. Physics, Vol. 35, No. 6, p. 2259, Dec. 1961.
9. O. S. Heavens, Optical Properties of Thin Solid Films, Butterworths, 1955.

THE CALIBRATION AT THE NATIONAL BUREAU OF STANDARDS OF MASS STANDARDS FOR ULTRAMICROANALYSIS

L. B. Macurdy
Mass and Scale Section
National Bureau of Standards
Washington, D.C.

ABSTRACT

Measurements may be made with a precision of tenths or hundredths of a microgram. Small weights suffer from an unfavorable surface-to-mass ratio and should be checked from time to time. This may be done by regular procedures used in weight calibration, provided a suitable selection of nominal weight values exist in each set.

*   *   *

This paper discusses the methods employed by the National Bureau of Standards Mass Laboratory for the calibration of weights of high accuracy as a service regularly offered, and their usefulness in scientific research. With some ultramicrobalances now available commercially, measurements may be made with a precision of tenths or hundredths of a microgram or perhaps to somewhat better than a hundredth of a microgram.

Such instruments may have an on-scale range of from several milligrams down to perhaps less than a milligram for the more sensitive instruments. Small weights of denominations within the on-scale range of an ultra-microbalance provide a means for direct calibration of the indications, if the accuracy of the weights is comparable to the precision of measurement. Such weights provide a means for estimation of and correction for the nonlinear limitations of the instrument. There are well established methods of weight calibration that are in regular use in the NBS Mass Laboratory which reduce the uncertainty in value of standard weights to less than the uncertainty associated with a single measurement with the balance used in the calibration. By the use of such methods weights can be calibrated on a particular ultra-microbalance with accuracy appropriate to the calibration of the on-scale range of that instrument. Other methods of calibration of ultramicrobalances are described in the literature.[1]

Small weights of the various laboratory classes intended for use with knife-edge balances are intended to be constant and accurate within appropriate fractions of much larger sample sizes than those used in ultra-microanalysis. They usually do not provide the accuracy required for the much smaller samples measured on ultramicrobalances. Requirements for suitable ultra-microbalance weights should specify such features as small surface-area-to-mass ratio, good surface finish, durable corrosion- and abrasion-resistant material, and dust-free case. Because small weights necessarily suffer from a relatively unfavorable surface-to-mass ratio, it is possible for the weights to change from a variety of causes. Accordingly, the values should be checked from time to time, and this may best be done by the methods recommended in this article.

The measurements required in weight calibration consist of individual comparisons of pairs of weights, or groups of weights of the same nominal value, which actually differ by relatively small amounts. The weights in a particular decade must comprise a series of denominations that permit comparison weighings of more combinations than the minimum number required to provide a solution for the values of the individual weights. An additional degree of freedom for the estimation of the standard deviation is provided by each weighing over the minimum required. All mass values are derived ultimately from the kilogram through a series of such intercomparisons. In the calibration of a set of weights, sufficient accuracy can usually be provided by use of reference standards whose mass has been previously determined by intercomparisons from the kilogram, through each intermediate decade, to the decade containing the standards. In the calibration of an intermediate decade, a summation of the weights must be included in the measurements at the next larger decade, and the weighings should include a summation from the next smaller decade, so that the smaller decade can be weighed in similar fashion. Two or more reference standards of denominations near the capacity of the balance should be included in the group for the largest decade. This provides assurance as to the proper identification of the reference standards, since the observed differences should agree with the accepted values for the standards within amounts that are consistent with the precision of measurement and the expected constancy of the standards.

A least squares solution leads to the "best" estimate of the values of the weights or groups of weights. When these adjusted values are substituted in the equations for the observations, residual errors ϵ_1, ϵ_2, ϵ_3, . . . are obtained that provide an estimate of the standard deviation of an individual measurement σ_a by the following relation:

$$\text{Estimate of } \sigma_a = s = \sqrt{\frac{\Sigma \epsilon^2}{f}}$$

where f is the number of degrees of freedom.

An estimate, σ_v, of the standard deviation of the value of a weight provided by the least squares solution is given by the relation below. The least squares solution for the value of a weight takes the form of the sum of a series of products, one product consisting of the appropriate fraction, D, of a summation of weights whose value was obtained by measurements in the next higher decade, the other products consisting of two terms, one an observation, the other an appropriate multiplier or coefficient c_i for the particular observation, determined so as to introduce the proper fraction of each observation into the "best estimate" of the value of the weight.

$$\sigma_v = [\Sigma c_i^2 \sigma_a^2 + D^2 \sigma_k^2]^{1/2}$$

where Σc_i is the sum of the coefficients of the observations in the least squares solution and σ_k is the standard deviation of the value of the summation of weights from the next higher decade. The proportional part of the uncertainty of the summation σ_k is combined as a random error with $\Sigma c_i^2 \sigma_a^2$, that is, by taking the square root of the sum of the squares of the two terms.

It is not always recognized that the least squares solution of a particular series of weight denominations can be reduced to a fixed series of operations so that the solution can be reduced to a repetitive procedure. Solutions to a variety of series of denominations are given in the literature,[2,3,4,6] and many of these solutions have been produced in the form of computation forms for use in the NBS Mass Laboratory. Copies of these forms and instructions for their use may be provided upon request. The estimates of the uncertainty in the determinations also are reduced to convenient equations.

The calibration of a set of weights by this method consists of a series of partial calibrations, one for each group of denominations in each decade. The estimates of the standard deviation of an individual measurement provide a good estimate of the weighing capability of the balance used and show the variation in precision with change in load under the particular laboratory conditions. Repetitions of the calibration for the set from time to time provide a basis for estimation of the constancy of the weights during use. These factors are important considerations leading to a valid estimate of the accuracy of measurement with an ultramicrobalance.

The question often arises as to what is meant by the accuracy of weights calibrated under the NBS weight calibration service. The accuracy of any measured value may be stated to be the extent to which the measured value is in agreement with the true value. Since the accuracy of mass measurement may extend through five or six figures, or more in special situations, an estimate of the accuracy is inconvenient simply because of the large number of digits involved. Since one or at most

two doubtful figures are included in certified values, it
is more convenient to think about the inaccuracy involved
in the doubtful figures. Three times the standard deviation is taken as the estimate of the maximum inaccuracy.
In the case of Class J weights[5] for use with ultramicrobalances, the estimate of the standard deviation is computed in detail for each calibrated value. Provided the
systematic error is relatively small compared to the
random errors, three times the standard deviation of the
value is a valid estimate of the maximum inaccuracy.
A detailed estimate of the maximum inaccuracy can be
made only after the measurements have been completed.
In the case of the classes of laboratory weights other
than Class J, the estimate is usually made in advance
of the measurements. Before such an advance estimate
can be made, the measurements must be brought under
strict control. Systematic effects larger than the random
errors of measurement must be avoided. For example,
proper weighing methods are required to eliminate the
effect of inequality of arms, and corrections must be
applied for the effect of air buoyancy.[6] Also, mistakes
such as numerical errors, misreadings, and incorrect
recording of weights, and the effects of abnormal behavior of the balances must be located or identified, and
must be eliminated from the measurements. The residual
errors computed in connection with the least squares
solutions are a valuable aid, since characteristic patterns
of the residuals are associated with errors in a particular
observation. In the NBS Mass Laboratory a check of
some sort is obtained on each balance indication that is
used in a weight calibration. More than one standard is
regularly used as the basis for the values, and the esti-

mates of the standard deviation of an individual measurement are inspected for conformity with established performance. Values of the standard deviation of a double-transposition weighing in the NBS Mass Laboratory are listed in Table I for individual weighings at various loads on several balances.

In our control of the measurement process we are continuing the use of methods developed by A. T. Peinkowsky, Chief of the Mass Section at NBS for many years,

TABLE I

Standard Deviation at NBS for Double-Transposition Weighings

Load	Standard deviation, μg	Balance
1 kg	7	Rueprecht
200 g	6.6	Ainsworth
100 g	5.2	Ainsworth
30 g	3.7	Ainsworth
2 g	1.4	Ainsworth
20 g	0.9	Corwin
2 g	0.5	Corwin
1 g	0.37	Keller
200 mg	0.33	Keller
100 mg	0.22	Keller
200 mg and less	0.02	Rodder (by substitution weighing)

during which time he established an enviable record for accuracy of measurement of mass.

The forms for the solution of sets of measurements for various combinations of weight denominations have been so devised that the standard deviation of the calibrated value will regularly be somewhat smaller than the standard deviation of the individual measurement. As a rough approximation, the estimate of standard deviation of a single measurement is usually slightly smaller by 10 or 20% than the maximum residual found in a series. We also know that the least squares adjusted value will have a standard deviation of from one-half to one-third of the value of the standard deviation for a single measurement. Therefore, the maximum residual of a series provides a fairly good estimate of the three-standard-deviation limit on the adjusted values. Alternatively, a glance at Table I gives a rough estimate of two or three times the standard deviation of the measured value at NBS. While the inaccuracy in Class J calibrations is within smaller absolute limits, as measured in micrograms, larger denominations in other classes may have higher percentage accuracy. Accordingly, any advance estimate should include some allowance for possible variability, unless such an effect is known to be a relatively small effect compared to the random errors of measurement.

The scientist who requests NBS weight calibration service must assume responsibility for continuing accuracy of value. The NBS contribution may be outlined as including specifications[5] that make possible constancy of value of the weights, the calibration of the values, and development of weight calibration methods. Responsibility of the scientist includes proper custody and care to insure that the essential feature of the standard weights is not lost through improper or unauthorized handling. The scientist should occasionally redetermine the values of NBS-certified weights by use of the same methods as those used at NBS.[3,4,6] This leads to a better knowledge of the capability of his weighing equipment for weighings to the highest accuracy. Through use of NBS weight calibration methods the scientist will in many cases find that he rarely will have to return the weights for recalibration. When NBS methods are not used, standards are frequently returned for recalibration because of doubt as to the continuing accuracy.

When the weight calibration service with accuracy to 0.1 or 0.2 μg was first offered, the Bureau found that no one had weights sufficiently constant to justify making the measurements. To get the service started several sets of Class J weights were made in the NBS Mass Laboratory and provided to those who needed the service. The balances used for these calibrations were selected assay-type knife-edge balances. An individual measurement on these balances had a standard deviation of 0.221 μg. The maximum uncertainty in the certified values was from 0.16 to 0.22 μg.

Current efforts in the NBS Mass Laboratory include a study of methods of cleaning weights of various materials as a basis for improvement in precision and constancy in value. We are studying a possible improvement in two-piece weight design. Studies on weighing instruments include a study of knife-edges and bearings of various materials and work on four new models of balances with capacities of 50 lb, 6 lb, 1 kg, and 2 g. The

2-g-capacity balance is a fused-silica balance and is intended to provide increased precision for the subdivision of 5-g summations of our National Reference Standards of mass. We also have recently put into service a quartz microbalance to provide improved accuracy in the calibration of weights of small denominations.

The methods of weight calibration recommended here will provide accuracy in value of the smaller weights of a set such that three times the standard deviation of the value given for the weight will not be larger than the standard deviation of a single measurement on the balance. Accordingly, these methods can be used to provide appropriate accuracy for any weighing device no matter how precise the instrument may be. These methods are recommended to balance manufacturers as a means of describing the performance of their balances in a precise and meaningful way. The methods are necessary for the scientist who wishes to assess the maximum capabilities of weighing instruments. The most useful standards for such an assessment would be several weights of denominations near the capacity of the ultramicrobalance, and several small weights within the on-scale range. The calibration of standard weights in such groups would provide the basis for the calibration of weights to the maximum accuracy required for a particular balance. The greater percentage accuracy of the larger weights near the capacity of the ultramicrobalance can be carried forward in proportion to smaller masses or mass differences within the limitations of the readability of the weighing instrument.

REFERENCES

1. Paul L. Kirk, Quantitative Ultramicroanalysis, John Wiley & Sons, New York, 1950.

2. John F. Hayford, "On the Least Square Adjustment of Weighings," Report for 1892, Appendix No. 10. U.S. Coast and Geodetic Survey, Washington, D.C.

3. M. J. Rene Benoit, L'étalonnage des Séries de Poids, Vol. XIII, Travaux et Memoires du Bureau International des Poids et Mesures, 1907. Translated from the French, June, 1960. Standardization of Sets of Weights by J. R. Benoit Library Convair (Astronautics) Division of General Dynamics Corporation, No. AE60-0003-10.

4. H. E. Almer, L. B. Macurdy, H. S. Peiser, and E. A. Weck, "Weight Calibration Schemes for Two-Knife Direct-Reading Balances," Engineering and Instrumentation, Section C, J. Research Nat. Bur. Standards, in press.

5. NBS Circular 547. Precision Laboratory Standards of Mass and Laboratory Weights. (Also included in NBS Handbook 77, Volume III.)

6. NBS Circular No. 3, Design and Test of Standards of Mass. (Also included in NBS Handbook 77, Vol. III.)

INDEX